Ketson Patrick de Medeiros Freitas
Genilson Pereira Santana

Generation of Electric Energy

Ketson Patrick de Medeiros Freitas
Genilson Pereira Santana

Generation of Electric Energy

From Urban Solid Waste in Isolated Systems of Amazon

ScienciaScripts

Imprint

Any brand names and product names mentioned in this book are subject to trademark, brand or patent protection and are trademarks or registered trademarks of their respective holders. The use of brand names, product names, common names, trade names, product descriptions etc. even without a particular marking in this work is in no way to be construed to mean that such names may be regarded as unrestricted in respect of trademark and brand protection legislation and could thus be used by anyone.

Cover image: www.ingimage.com

This book is a translation from the original published under ISBN 978-620-4-19267-3.

Publisher:
Sciencia Scripts
is a trademark of
Dodo Books Indian Ocean Ltd., member of the OmniScriptum S.R.L Publishing group
str. A.Russo 15, of. 61, Chisinau-2068, Republic of Moldova Europe
Printed at: see last page
ISBN: 978-620-4-03019-7

SUMMARY

The Isolated Systems - SISOL, which in the context of the electric grid are those localities not connected to the National Interconnected System - SIN, are mostly concentrated in the northern region of the country, more specifically in the state of Amazonas. These locations are affected by several problems, such as: energy deficit, high energy prices and large emissions of pollutants. At the same time, the wrong destination of Urban Solid Waste - USW in the state of Amazonas is a still existing problem. Given the above, the present study aimed to analyze the feasibility of generating electric power, from MSW, in SISOL Amazonas. To this effect, the SUW treatment technology was initially selected, with energy recovery, to be used in the analyses. This selection involved diverse parameters such as: costs, revenues, logistics, emissions and even social parameters. The selected technology was pyrolysis. SISOL of the state of Amazonas were also characterized, raising and estimating various data such as: population, income *per capita*, electricity consumption, MSW generation *per capita*, gravimetric composition, among others. Once selected the technology to be analysed and determined the data pertaining to each SISOL, the economic analysis was conducted in the certainty condition of the project. The implementation of the project proved to be economically unfeasible under the current conditions. However, it is noteworthy that these conditions involve parameters that were greatly impacted by the Covid-19 pandemic, such as the sale price of electricity and the dollar rate. Therefore, the sensitivity analysis was performed, in which three scenarios were evaluated, varying three parameters: sale price of electricity, dollar exchange rate and taxes. Of the three scenarios, the scenario that varied the three parameters simultaneously presented an economic viability of the project in 49.5% of the Amazon SISOL.

Keywords: Isolated Systems; Pyrolysis; Amazonas; Power Generation.

SUMMARY

1 ISOLATED ELECTRIC POWER SYSTEMS

In the context of electric energy, Isolated Systems (SISOL) are locations not connected to the national transmission grid, that is, to the SIN (National Interconnected System). This lack of interconnection occurs for technical or economic reasons. Thus, most of these systems are supplied electrically by local generation, usually based on diesel oil generators. (CCEE, 2017)

According to ONS (2019), there are currently 235 SISOL in Brazil, located mainly in the northern region, comprising the states of Acre, Amazonas, Amapá, Mato Grosso, Pará, Rondônia, Roraima and the island of Fernando de Noronha, which belongs to the state of Pernambuco. Figure 1 presents the geographical distribution of SISOL, as well as the total energy load in these systems, by state, forecast for the year 2020.

Figure 1- Geographical distribution and total expected energy load in 2020 of SISOL

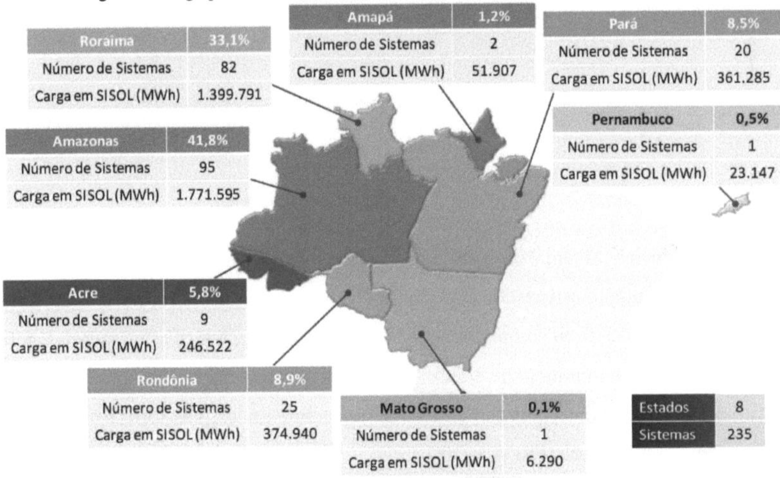

Source: ONS, 2019.

Observing Figure 1, it is possible to note the concentration of SISOL in the northern region, especially in Amazonas, which has the largest number of systems (95) and consequently the highest total energy load forecast for 2020.

1.1 SISOL in the state of Amazonas

It is valid to note the variety of the 95 SISOL located in the state of Amazonas, since they range from small localities, such as Carvoeiro, which has a verified demand of 58 kW in 2018, to larger cities such as Itacoatiara, with a verified demand of 31,680 kW (EPE, 2019). Figure 2 presents the geographical distribution of SISOL in the state of Amazonas.

Figure 2- Geographic distribution of SISOL in the state of Amazonas

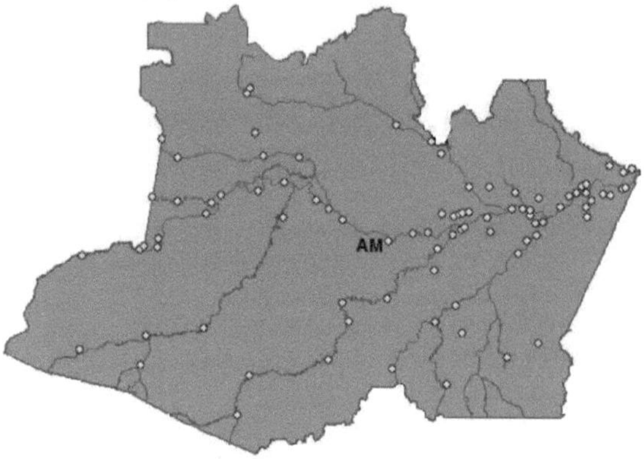

Source: EPE, 2019.

The map in Figure 2 shows that most of the localities are located along the rivers, which represent the main transportation route in the Northern region. The supply of electric energy in these SISOL depends on a complex logistics of fuel supply, which often suffers interruptions in times of drought, limiting the draft of the vessels. Furthermore, in many cases, the localities do not have appropriate port and airport infrastructure, as well as satisfactory transport and communication services (FROTA, 2004).

In Amazonas, despite having SISOL supplied by natural gas (Anamã, Anori, Caapiranga and Codajás) and also a location partially supplied by wood chips biomass (Itacoatiara), diesel engines have proven to be a solution of easy installation, maintenance and operation. Although it presents greater environmental impacts and complex logistics of fuel supply, with very high operating costs. As an example, while in SIN auctions the electric energy is traded at values of around R$200/MWh, in SISOL this value can reach R$1,600/MWh (PONTE, 2019).

The Ministry of Mines and Energy (MME), through the Planning of Assistance to Isolated Systems - Horizon 2024 (2019), conducted a survey and estimate of load and energy balance and demand for all SISOLs for the years 2019 to 2024. Annex A presents the data collected for the 95 SISOLs in the state of Amazonas.

By means of the demand balance information present in Annex A, it is also possible to verify the localities that will present energy deficit, as well as the year this shortage will start. Table 1 shows the locations with deficit forecast in the demand balance in 2024, the year the deficit starts, the forecast for interconnection to SIN and the types of fuel used.

Table 1- SISOL of Amazonas with forecast deficit in the demand balance

Location	Fuel Used	Demand Balance in 2024 (kW)	Deficit Starting Year	Forecast of Interconnection to SIN
Anamã	Natural Gas	-3664	2021	-
Anori	Natural Gas	-4664	2024	-
Caapiranga	Natural Gas	-5668	2020	-
Caiambé	Diesel Oil	-33	2023	-
Codajás	Natural Gas	-7414	2021	-
Humaitá	Diesel Oil	-20100	2023	mar/2023
Itacoatiara	Diesel Oil and Cavaco	-42290	2020	aug/2020
Itapiranga	Diesel Oil	-2858	2021	Dec/2020
Parintins	Diesel Oil	-30958	2020	mar/2024
Rio Preto da Eva	Diesel Oil	-10367	2021	Dec/2021
Silves	Diesel Oil	-1764	2021	Dec/2020

Source: Adapted from EPE, 2019.

Given these various localities with energy deficit, it is already possible to justify the search for new energy alternatives for SISOL. This search for alternative energy sources can also be corroborated when analyzed the emissions of pollutants caused by local energy generation in SISOL.

1.2 Pollutant Emissions

As mentioned, local energy generation in SISOL is predominantly by means of thermoelectric plants powered by diesel oil. This type of generation raises many questions about the emissions of pollutants (particulate matter) and greenhouse gases, which becomes more relevant because of the location of SISOL, in the middle of the Amazon rainforest (PONTE, 2019).

Several efforts have been made to connect the SISOL to the SIN, seeking to reduce the need for local generation and, consequently, emissions. However, the large extension of electrical networks requires a high investment cost, and in several cases, presents technical and economic restrictions (RIBEIRO, 2012).

The MME also carried out a survey of the estimated amounts of energy generated and carbon dioxide (CO_2) emitted for the year 2020. Table 2 presents the emission information, in millions of tons of CO_2 equivalent, in the energy generation estimated for 2020, by type of energy source.

Table 2- Amount of CO_2 emitted from power generation in SISOL in 2020

Source	Energy Generated (MWh)	Fuel Consumption		Emissions ($MtCO_2eq/ano$)
Diesel Oil	4.228.396	1.175.494	$m^3/year$	2,764
Natural Gas	165.232	44.612.585	$m^3/year$	0,092
Biomass	52.560	142.800	$ton/year$	0,004
Photovoltaic	6.555	-		0
PHC	54.844	-		0
Total	4.507.587			2,871

Source: EPE, 2019.

Using the total values obtained, of energy generated (4.507.587 MWh) and estimated emissions (2,781 $MtCO_2eq$), it is estimated that in the year 2020 an emission rate of 0.637 tCO_2eq/MWh. For comparison purposes, according to the MME itself, in 2017, the SIN presented an emissions intensity indicator of 0.090 tCO_2eq/MWh that is, more than seven times lower than that estimated for SISOL in 2020.

Therefore, the use of alternative sources for local energy generation in SISOL is an excellent alternative not only for localities with energy deficit, but also for those that have their energy matrix dependent on diesel oil, which are the majority.

2 ENERGY GENERATION FROM MUNICIPAL SOLID WASTE

The treatment of municipal solid waste (MSW) with energy recovery, is something positive both as an alternative source of local energy generation in SISOL, and as a way to circumvent the problem of final disposal of MSW in the national and local context.

2.1 Basic SUW Concepts

SUWs comprise those that are produced by the numerous activities that take place in areas where human settlements take place. It comprises waste of various origins (residential, commercial, agricultural, etc.), and its final disposal is usually the responsibility of the municipal authorities (ZANTA; FERREIRA, 2003)

This definition sizes up the complexity of the concept and meets the definitions of other works (IPT, 2000; MANCINI, 1999) in which the term RSU is often used as a synonym for garbage, and corresponds to any solid material arising from the daily activities of man in society, whose producer or owner does not consider it as something of sufficient value to preserve it.

SUWs may be classified according to their physical nature, their chemical composition, their degree of degradability, their potential risks to the environment and public health (periculosity), or even according to their origin or nature. The details of each type of classification are presented in APPENDIX A.

As for the characterization of the RSU, it consists in determining quanti and qualitatively its physical-chemical and biological aspects. Thus allowing the verification of the materials present in the waste generated, enabling inferences about the feasibility of the implementation of selective collection, the necessary human resources, dimensions of facilities and proper disposal for the waste (FERNANDO; LIMA, 2012).

SUW can be characterized by biological, chemical or physical characteristics. Details of each type of characterization are presented in APPENDIX B. It is relevant to cut out from the characterizations presented, the concepts of

(i)   Calorific Power: Determines the potential capacity of a material to give off a certain amount of heat when subjected to burning.

(ii)   Gravimetric Composition: indicates the percentage of each component in relation to the total weight of the waste sample analyzed, allowing the analysis of its total calorific value.

(iii)   *Per capita* generation: relates the amount of MSW generated daily and the number of inhabitants in a given region.

2.2 Integrated Management and USW Management

First of all, it is necessary to understand the difference between integrated management and SUW management. The former aims to develop guidelines in order to discipline actions, considering the various aspects involved (environmental, cultural, economic, political, social, etc.), so that the measures adopted are sustainable. As for the management, are precisely the actions disciplined by the integrated management. Thus, the management is part of the management, and can be understood as the steps to be performed (LOPEZ, 2007)

2.2.1   SUW Management in the Brazilian Context

APPENDIX C addresses SUW management within the international context, in diverse localities such as: the European Union, the United States of America and Japan. In Brazil, in 2010, the National Solid Waste Policy (PNRS) was created, which is a Federal Law (No. 12.305/10) that aims at the integrated management of SUW in Brazil. The PNRS covers various types of waste (domestic, industrial, hazardous, etc.), with the exception of radioactive waste, which has its own legislation.

It is important to highlight that for the implementation of the PNRS, the elaboration of Solid Waste Plans is foreseen for each administrative sphere of government, that is, national, state, regional, inter-municipal and municipal plans. A characteristic of the municipal plan, for example, is that it is a necessary requirement for municipalities to have access to federal resources.

The plans are basically strategic documents that determine, for an area of coverage, goals guided by the PNRS objectives. One of the main objectives of the PNRS is to reduce the amount of SUW that is sent to landfills and dumps. This objective led to the establishment of deadlines (the first being set until 2014) for municipalities to extinguish landfills. Several extensions have occurred and the total objective has not yet been (and is far from being) met.

2.2.2 *SUW Management in the Context of the State of Amazonas*

The state of Amazonas is the largest Brazilian state in territorial extension, this dimension associated with the dynamics of urbanization and crises in public management, favor the noncompliance with regulations, inspections and implementation of effective systems for environmental issues (ARAÚJO; SCHOR, 2011).

The survey released by the Ministry of Environment - MMA (2015), points out that 96.8% of the municipalities of the state of Amazonas have the Municipal Plan for Integrated Solid Waste Management (PMGIRS). Despite appearing to be a favorable information, it is worth remembering that this plan is a necessary condition for municipalities to have access to Union resources, which causes these plans, many times, to be made in a hurry and are hardly fulfilled.

Corroborating the idea that most PMGIRS are merely an existing but inapplicable document, the same survey published by the MMA in 2015 collected data as to the final disposal of SUW in each municipality. Of the Amazonian municipalities, 90.3% dispose of SUW in dumps and the remainder in controlled landfills. This demonstrates a total nonconformity with PNRS objectives that ought to guide municipal plans. It is noteworthy that no municipality has a landfill that meets specific conditions to be considered a sanitary landfill.

Given this fact, the search for alternatives that reduce SUW disposal at dumps (which is already theoretically prohibited) and also at landfills (which is more adequate for waste that cannot be treated or reused) is essential.

2.3 Disposal of MSW

The term destination translates mainly into treatment and final disposal of SUW. According to Abreu (2019), the main forms of SUW treatment and disposal are: recycling, composting, landfill, biodigestion, incineration and pyrolysis.

2.3.1 *Recycling*

According to Lajolo (2003) and Teixeira (2000), recycling is a process that consists of a set of interconnected operations, performed by different economic agents, in which the material, after its use, returns to the production cycle, whether that of its origin or any other.

This reintroduction to the productive process, in addition to recovering and reducing the quantity of SUW, significantly saves the amount of energy spent in industrial production. Basically, it consists of a few steps, such as the separation, collection and sorting of MSW according to type, classification and pressing. Subsequently, some kind of processing takes place and finally, recycling itself takes place and the material is reintroduced to the production process (CALDERONI, 2003; CUNHA, 2002; RENÓ et. al., 2002).

2.3.2 *Composting*

Composting is a SUW treatment technique in which organic matter, under appropriate conditions of temperature, humidity and aeration, is transformed into a stable product known as organic compost, which has soil conditioning properties and is widely used in agriculture.

In the process itself, organic materials present in SUW are separated from inert materials (aluminium, glass and plastic) and taken to appropriate locations (composting yards, silos or rotating drums), where they will undergo a controlled degradation process, aiming at the production of compost (PAVAN, 2010).

The separation that occurs in the process helps in the quality of the compost. In view of this, composting depends on a selective collection, occurring, generally, integrated with a recycling process. The compost formed is used in market segments such as: agriculture, gardening, landscaping, soil coverage and reforestation.

2.3.3 *Sanitary Landfill*

The sanitary landfill is a technique that consists of the disposal of MSW in the soil, based on engineering criteria and operational standards, ensuring the protection of both the environment and public health. In this process, the disposed SUWs are compacted and covered with soil in the form of daily cells, forming layers of waste. It is essential the adoption of engineering principles for treatment and collection of percolated liquids generated, in addition to a bottom waterproofing(MUTZ et al., 2017).

It is important to differentiate between sanitary landfills and controlled landfills and dumps. In dumps, or waste dumps, SUW is discharged onto the ground, without technical criteria and environmental protection measures or public health, becoming a place of generation of disease vectors and having its leachate infiltrated into the soil. As for controlled landfills, despite the existence of measures to minimize impacts, the base sealing is not observed, often also causing the infiltration of its leachate (ROSA et al., 2017).

Integrated to the sanitary landfill, one may observe biogas collection, burning and processing systems, being an alternative energy use from SUWs.

2.3.4 Anaerobic Digestion

Anaerobic digestion is a biochemical process in which organic matter is degraded and stabilized in the absence of air, through the action of microorganisms, resulting in a product called biogas. Biogas is rich in energy, because it has a high percentage of methane. The process also generates an effluent called biofertilizer, which is widely used as a soil conditioner in agriculture (MAYER, 2016).

Because it is a process focused on the organic part of SUWs, as in composting, biodigestion occurs in an integrated manner with selective collection. It is also worth noting that there are techniques for capturing methane gas in landfills, but they have a lower production efficiency than the production in biodigesters. The biogas produced can be burned directly or used as fuel for internal combustion engines and gas turbines in power generation (PEDROXA et al., 2017).

2.3.5 Incineration

Incineration is a process that reduces the weight, toxicity and volume of MSW, through controlled combustion at high temperatures. The products of the process are ash, water, slag (such as ferrous metals, glass and stones) and gases such as CO_2, SO_2, N_2 oxygen from excess air and inert gases from the air and from the MSW itself. In cases of incomplete combustion (which is not ideal), carbon monoxide, soot, dioxins, furans and nitrogen dissociates may still be observed (LIMA, 1991).

Incinerators require high values of investment and higher operating costs than the other methods presented. This is due to the fact that they are more complex units, with a high degree of automation, many control devices and the need for a more qualified workforce. It is worth noting that, generally, the energy released in the burning is used to generate steam and electricity.

2.3.6 *Pyrolysis*

Pyrolysis is a thermal treatment process that promotes the decomposition of SUW in an oxygen-free atmosphere. In this process, MSW is sent to a reactor, usually a furnace without the presence of oxygen, where it is subjected to high temperatures for a certain period of time. The process results in high quality fuels, which can be gaseous, liquid and solid, depending on the purpose of the plant (ABREU et. al., 2019).

Also according to Abreu (2019), pyrolysis processes generally rely on the treatment of gaseous and liquid effluents, generating practically no environmental liabilities. Like incineration, pyrolysis significantly reduces the mass and volume of MSW. Given that this is a relatively new technology, it is a method that is few examples of implementation on a large operational scale.Logically, fuels generated from SUWs in a pyrolysis process can and are also used to generate energy.

2.4 MSW Treatment with Energy Recovery

According to Pedroxa (2017), the main treatment and disposal technologies for the conversion of MSW into energy, more specifically into electrical energy, are: anaerobic digestion, incineration and pyrolysis.

2.4.1 *Anaerobic Digestion with Energy Recovery*

According to Verma (2002), anaerobic digestion is a consequence of several metabolic interactions between various groups of microorganisms. The process occurs, basically, in three stages: hydrolysis, acidogenesis and methanogenesis. First, a group of microorganisms secrete enzymes, hydrolyze polymeric materials to monomers such as glucose and amino acids. Next, another group, converts the materials formed into hydrogen and acetic acid. And finally, other bacteria are responsible for converting the hydrogen, carbon dioxide and acetate into methane.

Anaerobic digestion can be divided into two phases: mechanical phase and biological phase. And its systems can also be classified as to temperature, as to the solid content and as to the type of system used. The details of both the phases and the classifications of the systems are presented in APPENDIX D.

According to Costa (2010), anaerobic digestion is not a single technology, but a combination of technological processes that aims to achieve the desired objective. These objectives are usually: increased recovery of recyclable materials, production of biofertilizers, production of stabilized material for landfill, production of heat and electricity and production of fuel derived from waste.

Given that the biodigestion treatment system is focused on the organic fraction of SUWs, one notices and is further reinforced by the EPE (2008) study, that recycling is an alternative that compares very well with the generation of electrical power as of the anaerobic digestion of SUWs.

Anaerobic digestion occurs naturally in various environments, such as: swamps; sediments of rivers, lakes and seas; coal mines; digestive tract of animals and landfills (AMARAL, 2004). Thus, several technologies for capturing methane in landfills have been developed and implemented. Another way to use the anaerobic digestion of MSW is to perform it in a controlled manner in reactors, generally known as biodigesters.

According to several studies (AMARAL, 2004; LEITE, 2016; SOUZA et al., 2012), the most appropriate system for anaerobic digestion of MSW depends on the characteristics of the waste, the available area, financial and operational resources, the importance of energy generation for the system, pollution prevention, and several other factors. All of these factors make anaerobic digestion of SUWs in reactors more attractive than landfill capture. This is due to the possibility of, in reactors, potentiating biogas production, having better operational control of the process, greater ease in capturing biogas and shorter residence times when compared to landfills.

2.4.1.1 Biodigester Models

Biodigesters are normally a physical structure with a chamber where the process of degradation of organic matter occurs. This structure can be cylindrical, vertical and superficial, and may or may not be above ground, accompanied by a bell (also known as a gasometer) where the biogas formed is accumulated (PINTO, 2008).

Frigo (2015) in his article on models and applications of biodigesters, deepens the knowledge in different models of biodigesters, such as: Chinese model, Indian model, Canadian model and batch model. As it can be used in small and large projects, presents a greater production of biogas and is the most widespread in Brazil, the model considered in this work is the Canadian type biodigester. The other models are discussed in detail in APPENDIX E.

The Canadian model biodigester is characterized by being horizontal, with a masonry fermentation chamber wider than the depth, which promotes a greater area of exposure to the sun, increasing the production of biogas and even avoiding clogging (CASTANHO; ARRUDA, 2002).

Another feature is the existence of an upper blanket fixed on a ditch of water that surrounds the base, forming a different storage hood, where it is also located, the register for gas output. The fermentation chamber is also covered with a plastic canvas (PEREIRA; DEMARCHI; BUDIÑO, 2009).

It is necessary an installation site that provides the lowest risk of occurrence of holes in the upper blanket that may cause leakage of biogas. This type of biodigester is the most widespread in Brazil, and can be used in both small and large facilities (FRIGO et al., 2015).

Figure 3 shows the configuration of a biodigester of the Canadian model. The PVC tarpaulin fixed over a water ditch surrounding the base can be seen, forming the gasometer of the system. The fermentation chamber is also lined with a waterproofing material.

Figure 3- Biodigester: Canadian Model

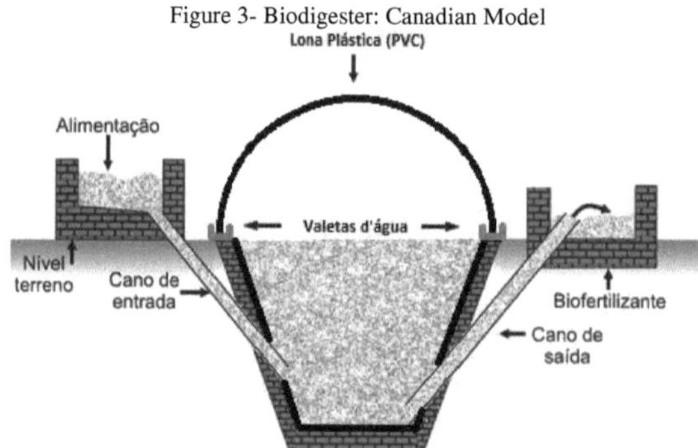

Source: Adapted from Perlingeiro, 2014.

2.4.1.2 Advantages of Anaerobic Digestion

Some of the advantages of an anaerobic digestion system, in general, are: reduction in volume and moisture of MSW for final disposal, reduction in the amount of organic material disposed of in landfills, reduction in the formation of gases that generate the greenhouse effect, generation of important agricultural inputs, decentralization of energy generation, promotion of an integrated management of MSW, among others (LEITE, 2016)

2.4.1.3 Disadvantages of Anaerobic Digestion

As disadvantages of anaerobic digestion can be cited, for example, the formation of sulfides by the reduction of sulfates, which ends up generating hydrogen sulfide gas due to the non-generation of methane from oxidized matter. This gas has an unpleasant odor and oxidizes metallic components. Some other disadvantages are: complex biochemistry and microbiology of the process, sensitivity of bacteria to various compounds, possibility of delay in starting the process, low removal of phosphorus and nitrogen, among others (Meyer, 2016).

2.5 Incineration

Incineration may be defined as a process that reduces SUW weight and volume via controlled combustion (GRIPP, 1998). Menezes (2000) further adds by presenting that in the past, incineration's prime objective was that of reducing mass and volume and currently, mechanisms are incorporated to take advantage of thermal energy via combustion process energy derived from SUW's calorific power.

According to Machado (2015), incineration is the most widespread high temperature thermal treatment process, with a high number of units in commercial operation around the world. This is especially true in countries with little available landfill area, such as Japan, Switzerland and Sweden. It is estimated to reduce between 12 and 30% of the mass and 4 and 10% of the volume of the original waste.

Basically, incineration systems consist of chambers where controlled combustion of MSW occurs, presenting gases, water, ash and slag as remainders. And additionally, some mechanism of energy utilization, usually steam thermal machines.

Steam thermal machines use the energy from SUW combustion to a working fluid, usually water, generating steam and converting the high energy of the fluid into mechanical work or heat for secondary heating processes. This kind of system has been responsible for a large portion of the electrical energy produced worldwide (HENRIQUES, 2004).

These energy recovery systems generally consist of: boilers (where steam is produced and accumulated), turbines (which convert thermodynamic energy into mechanical work), generators (to convert mechanical energy into electricity), condensers (to return steam to the liquid state), and pumping systems. These features are recurrent in the various models of incinerators.

2.5.1 Models of Incineration Systems

There are a variety of incineration systems. One division that is generally accepted is the categorisation into grate and kiln systems. The (i) grate type systems are suitable for large, irregular waste held on a movable or stationary grate, allowing air to penetrate through the grate and pass through the waste. Furnace type incinerators are diverse and include those of the type: Rotary Furnace, Multiple Fixed Chamber, Fluidized Bed, Liquid Injection and Plasma type. The different types of furnace incinerators are discussed in APPENDIX F.

The grate type incinerators are the most suitable for the treatment of MSW, being even the most used equipment for incineration of MSW. Basically, in this model, SUWs are burnt on metal grids that allow air to circulate under, over and through the waste. Evidently, this type of incinerator depends on the waste remaining on the grate and not falling into the ash pit before being burned. Movable grate incinerators are widely used for municipal waste and rarely for process waste (LANÇA, 2008)

Figure 4 shows the thermal process, already with energy utilization, using a grate type incineration system. It can be seen that the MSW are discharged into the pit (1), from where they are forwarded, through grabs, to the hopper (2). The waste in the hopper feeds the incinerator (3) of the grate type, which in turn provides heat, through combustion, to the boiler (4). In the boiler, steam is produced and conducted to the turbine-generator system, for electric power generation. The remaining ash on the grate passes through electromagnetic separators (5). And the gases resulting from combustion are routed to treatment systems (6 and 7) for removal of pollutants, being subsequently released, through the chimney (8), to the environment.

Figure 4- Thermal treatment process with grate incinerator

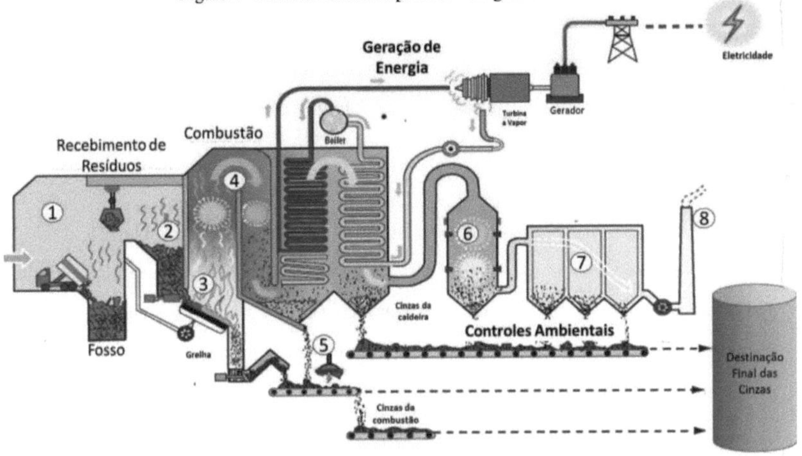

Source: Leite, 2016

2.5.2 Incineration Advantages

In general, one of the advantages of incineration is the ability to considerably reduce the volume and mass of SUW to be disposed of in landfills. Some authors also consider as an advantage, the fact that there is generally no need for SUW separation. Generally, incineration presents a higher energy yield than anaerobic digestion. And the incineration process is also more advantageous in relation to the destruction of health services waste (LEITE, 2016).

2.5.3 Disadvantages of Incineration

The main disadvantage that is raised by most people, is the effluents of the incineration process. In addition to the common products of combustion, that is, water vapour and carbon dioxide, other pollutants are also produced, such as dioxins and furans, which are persistent organic pollutants, many of which are toxic carcinogenic elements that accumulate in the fatty tissue of mammals (IDEC, 2005).

In Brazil, incineration plants have been deactivated, mainly due to the precariousness of the installations, and there are controversies about the acceptable levels of exposure to dioxins, which in most cases contaminate by ingestion, but can also contaminate by inhalation (CAIXETA, 2005).

Finally, a widely debated characteristic is the issue of the incineration process competing with the recycling alternative. Given that incineration, in addition to not requiring prior separation, is still favoured, energetically speaking, with the calorific power of materials such as paper and plastic which are the components that provide the greatest recycling benefit.

2.6 Pyrolysis

Pyrolysis consists in the decomposition of organic matter heated in the total (or partial) absence of atmospheric oxygen or other oxidizing agent. This heating is controlled by temperature ranges, providing the necessary energy for the breakdown of macromolecules present in the waste biomass (DINIZ, 2005).

The pyrolysis processes occur through complex chemical reactions, which depend qualitatively and quantitatively on the reaction conditions, in addition to the characteristics of the waste itself. Some of the main variables observed in a pyrolytic operation are: heating rate of the reactor, pressure in the reactor, reaction temperature and residence time of both solid phases and vapors (PEDROXA et al., 2017).

The main products of pyrolysis are the formation of coal, bio-oil and fuel gas. These products can be used to generate heat and electricity or be upgraded to be used as fuel or chemicals. Depending on the conditions of the reactor, the production of one of these products can be maximized, and it is precisely these conditions that determine the different models and types of pyrolysis (MOTA et al., 2015)

2.6.1 Types of Pyrolysis

According to Mota (2015), a well accepted division for the types of pyrolysis processes, is the classification into: (i) Slow Pyrolysis, (ii) Fast Pyrolysis and (iii) Ultrafast Pyrolysis. Slow pyrolysis is the less complex type and consequently with a more possible applicability, being the one considered in this work. Both fast and ultrafast pyrolysis are discussed in APPENDIX G.

The slow pyrolysis, also known as conventional pyrolysis and even carbonization, is characterized by small heating rates and a maximum temperature range around 600 to 800°C. In addition to a residence time of biomass in the reactor between 5 and 30 minutes, or even hours (MOTA et al., 2015)

The main product of slow pyrolysis is coal, which has a much higher energy density than the original material and burns at much higher temperatures. In addition, in smaller quantities, fuel gas, bio-oil and pyrolenoic acid are produced (PEDROXA et al., 2017).

Figure 5 shows a MSW treatment system with energy recovery, using the slow pyrolysis process. Basically, SUW feed the hopper (1), being sent through the conveyor belt (2) to the carbonization oven (3). In the oven, the waste is heated without the presence of oxygen, which promotes the decomposition of organic matter without combustion.

The gases formed are channeled (4) to the distiller (5), where they are liquefied generating the pyrolene extract (a byproduct widely used in the chemical industry and agriculture). After the distillator, the remaining gases still pass through a filter (6) where the solid particles are separated, releasing only water vapor through the stack (7).

The MSW, transformed into coal, is sent to the separator (8). In the separator are removed for recycling, through the conveyor belt (9), the non-carbonized materials, such as aluminum, glass, iron, copper, etc. Part of the coal (about 10%) is used as fuel for the furnace and the rest is sent to the steam turbine system (10), to generate electricity through the engine, generator and transformer.

Another interesting point is the fact that the ash, both from the furnace (11) and the boiler (12), are directed into the furnace, being part of the decomposition process. In addition, the gases resulting from burning coal, both in the furnace and in the boiler, are channeled and directed to the distiller and filter. This makes that, in theory, the emission of this system is reduced to water vapor.

Figure 5- Slow Pyrolysis Process with Energy Recovery

Source: Adapted from Usitrar, 2017.

2.6.2 *Pyrolysis Advantages*

The use of the pyrolysis process in the treatment of MSW presents advantages such as the significant reduction in the volume of MSW and also like incineration, it can be used for the satisfactory disposal of health service waste. Minimum generation of environmental liabilities. Possibility of generating direct and safer and healthier jobs for Gcolaboradores. Use of a small space for installation of the enterprise (ABREU et. al., 2019).

2.6.3 *Disadvantages of Pyrolysis*

Because it is an innovative technology, there is still a lot of resistance to its implementation on a large operational scale. Due to the high risk of air contamination, the costs of gaseous effluent treatment tend to be high. In addition, it generally requires a more qualified workforce, which can make the operational and maintenance costs higher than the others (ABREU; HENKES, 2019; MONTEIRO et al., 2001).

3 SELECTION OF THE MOST RELIABLE TECHNOLOGY FOR THE CASE STUDY

This chapter compares and ranks MSW treatment technologies with energy recovery based on different parameters such as: costs, revenues, emissions, logistics and social. In light of the ranking, one is able to select the most adequate technology for SISOL. The technology selected in this chapter will be used in the case study application, presented in subsequent chapters.

3.1 Selection Methodology

This selection utilised the technologies that generate energy as of SUW presented in chapter 2: Anaerobic Digestion, Incineration and Pyrolysis. Amongst the types of each technology, those most adequate for the treatment of SUW in small and medium sized cities were considered, i.e., the most efficient systems, but with the lowest possible cost and degree of complexity, which generally implies the most commonly used systems: Canadian model biodigesters, grate type incinerators and slow pyrolysis.

Regarding data and information, secondary data obtained from several published works were used. Because it often deals with issues with little information, as is the case of pyrolysis, and seeking to perform a diversified analysis and as close as possible to reality, it was used in most cases, the mathematical relationship of transitivity.

The transitivity relation is the one that is established between three (or more) elements, in such a way that if the first has relation with the second and this has relation with a third, then the first has relation with the third. A classic example is if there is the relations "$A>B$" and "$B>C$", then there is also the relation "$A>C$" (LIMA; NEVES, 2019).

As many studies presented data on technologies, but in different environments (amount of MSW, gravimetric composition, etc.), it was unfeasible to directly compare this information. Thus, we used the transitivity relationship to classify technologies.

Illustrating the approach to the relationship of transitivity in this study, one can observe as an example that some studies compared anaerobic digestion and incineration, while others addressed incineration and pyrolysis, thus, it was possible to establish a relationship and consequently a classification of the three technologies. It is worth noting that various studies were used, which diversifies the analyzed environments, minimizing the trend characteristics.

The PNRS states that any waste processing practice or technology to be licensed must be clean and environmentally friendly, economically viable and socially inclusive (MEDEIROS; CASTRO, 2015). In light of this, the selection conducted in this chapter was based on five different analysis parameters: (i) Costs, (ii) Revenues, (iii) Emissions, (iv) Logistics and (v) Social.

(i)   Costs: This is a quantitative parameter, and considers: (i) deployment costs (CAPEX) and (ii) operational costs (OPEX).

(ii)   Revenues: Basically, it is the sum of the multiplications between the quantity of products generated and their price estimates. This parameter is quantitative and considers: (i) electric energy revenue and (ii) revenue from other by-products.

(iii)   Emissions: This is a quali and quantitative parameter, considering: (i) the types of pollutants emitted and (ii) the weight of the emissions in a multi-criteria analysis.

(iv)   Logistics: This is a mostly qualitative parameter, observing factors such as size of facilities, destination of by-products, processing time, etc.

(v)   Social: It is already a qualitative parameter, observing issues such as competition with recycling, labor aspects, among others.

For each selection parameter, the technologies were evaluated and ranked. This classification, in the form of a *ranking,* allowed the adoption of a score to determine the selected technology through a quantitative analysis. It is worth remembering that the PNRS does not have a clear distinction of the importance of each analysis parameter, so all received the same weight in this selection.

3.2 Cost Analysis

In the cost analysis, both initial investment costs (CAPEX) and operating costs (OPEX) were taken into consideration. In each case, different works were analyzed, the necessary currency conversions were performed and the cost averages were calculated. Finally, CAPEX and OPEX were summed, and based on this result, technologies were ranked in the "costs" parameter.

3.2.1 *Implementation costs - CAPEX*

To determine CAPEX, the guides for decision makers of Mutz (2017) and CNI (2019) were considered. The former presents an estimate of costs in Euros, while the latter gathers and summarizes several works, presenting an overview of costs in Dollars. Table 3 brings together the information collected in both works, already converted and adapted to the format of Dollars per Capacity, i.e. US$/(tons per year), as well as the average of the values.

Table 3- CAPEX of MSW treatment technologies with energy recovery

Source	CAPEX (US$/ton/year)		
	Anaerobic Digestion	**Incineration**	**Pyrolysis**
CNI, 2019. [1]	334,00	723,50	574,00
Mutz et al., 2017. [2]	172,80	378,00	432,00
Average	**253,40**	**550,75**	**503,00**

Source: Adapted from Mutz (2017) and CNI (2019).

3.2.2 *Operating Costs - OPEX*

For the determination of OPEX, the works of Mutz (2017) and CNI (2019) were also used. The former presented specific operating costs, as well as the total operating cost (value that was considered in the present work). As for the study of CNI (2019), analogously to what was done in CAPEX, the average of the values presented for each information was used. Table 4 shows the OPEX data, in Dollars per year, in addition to the average of these.

Table 4- OPEX of MSW treatment technologies with energy recovery

Source	OPEX (US$/ton)		
	Anaerobic Digestion	**Incineration**	**Pyrolysis**
CNI, 2019.1	100,50	118,50	45,00
Mutz et al., 2017.[2]	30,24	87,48	81,00
Average	**65,37**	**102,99**	**63,00**

Source: Adapted from Mutz (2017) and CNI (2019).

---

[1] The average of the data gathered in the CNI guide (2019) was used.
[2] The values from Mutz's (2017) guide were adapted to the format "investment x capacity" and converted the values according to the following rate: 1.00 EUR = 1.08 USD.

3.2.3 Total Costs

As other costs, such as taxation for instance, tend to be similar or previously unmeasurable, the total cost considered in this study is given by the sum of the average CAPEX and OPEX values obtained for each technology. This information is presented in Table 5.

Table 5- Average total costs of MSW treatment technologies

Cost Type	Costs (US$/ton)		
	Anaerobic Digestion	Incineration	Pyrolysis
CAPEX	253,40	550,75	503,00
OPEX	65,37	102,99	63,00
Total	318,77	653,74	566,00

Source: Author, 2021.

3.2.4 Ranking on the basis of costs

The partial ranking is performed in each analysis parameter, and will serve as support for the overall ranking. This classification based on costs takes into consideration the average total costs obtained, ranking them from lowest to highest. The best ranked technology obtains 3 points, while the last ranked technology obtains 1 point. The intermediate technology receives 2 points. This scoring pattern is maintained in the other analysis parameters. And the points are used for a quantitative analysis in the overall ranking. Table 6 presents the partial ranking along with the scores obtained by each technology in the parameter "costs".

Table 6- Partial ranking of the technologies based on the "Costs" parameter

Parameter: Costs	Partial Ranking		
	Anaerobic Digestion	Incineration	Pyrolysis
Considered Value	318,77	653,74	566,00
Position	1º	3º	2º
Score	3 points	1 point	2 points

Source: Author, 2021.

3.3 Revenue Analysis

In the revenue analysis, we considered the revenue from the sale of electricity produced with the technologies and the revenue from the sale of by-products. We used studies that presented a paired comparison of the technologies, and through the transitive property of the relations obtained it was possible to partially classify the treatment technologies.

3.3.1 Revenues from the sale of electricity

To determine the relationship between technologies in terms of revenue from the sale of electric power produced, the studies of Bain & Company (2012) and Medeiros and Castro (2015) were utilized. The first presents the values, in reais per ton, referring to the anaerobic digestion and incineration technologies, using for this the energy efficiency of each plant and a sales value of 150 R$/MWh. The second study presents the energy efficiency of the incineration and pyrolysis technologies. Using the same sale value of the work of Bain & Company (2012), one could estimate revenues as of Medeiros and Castro's (2015) data. Table 7 gathers and presents the revenue values of both works.

Table 7- Revenues from the sale of electricity produced by the technologies

Source	Revenue (R$/ton)		
	Anaerobic Digestion	Incineration	Pyrolysis
Bain & Company, 2012.	37,50	100,00	-
Medeiros and Castro, 2015.	-	81,6	85,65

Source: Adapted from Brain & Company (2012) and Medeiros and Castro (2015).

3.3.2 *Revenues from the Sale of By-Products*

Also according to Bain & Company (2012), the generation of electricity is the only effective additional revenue from the incineration process. In relation to anaerobic digestion, one notices that revenue from the generation of electrical power is much lower than that from incineration (<40%); thus, revenue from non-mainstream products such as bio-fertilizer would not alter the placement of the technology.

Pyrolysis, in turn, which according to Medeiros and Castro (2015), already presents a higher main revenue than the others, in addition to having by-products, such as the pyrolytic extract, which even with a less relevant revenue, serves to further highlight the technology of the others.

Therefore, the revenue from the sale of by-products, in general, is not able to change the scenario of technology classification according to the revenue from the sale of electricity produced.

3.3.3 Revenue Ranking

The partial ranking based on costs was made through the transitive property of the relations presented. Due to the, in general, lower impact of by-product revenues in the total revenues, the relations between the revenues from electricity generation were used in this analysis. The technologies were classified from the highest to the lowest revenue. Table 8 presents both the relations obtained by each source and the general relation between the technologies, as well as the positioning and score in the partial ranking.

Table 8- Partial ranking of the technologies based on the "Revenue" parameter

Parameter: Revenues	Partial Ranking		
	Anaerobic Digestion	**Incineration**	**Pyrolysis**
Bain & Company, 2012. **Medeiros and Castro, 2015.**	Anaerobic Digestion < Incineration - - Incineration < Pyrolysis		
General Relationship	Anaerobic Digestion < Incineration < Pyrolysis		
Position	3º	2º	1º
Score	**1 point**	**2 points**	**3 points**

Source: Author, 2021.

3.4 Analysis Regarding Emissions

In the emissions analysis, the types of emissions of each treatment method were considered, in addition to a multicriteria analysis used to rank the technologies. Based on these qualitative and quantitative data it was possible to determine the ranking for the parameter "emissions".

3.4.1 Weights in Multicriteria Analysis

Marchezetti (2009) conducted an evaluation of technological alternatives for SUW treatment by means of a multi-criteria analysis method named AHP (Analytic Hierarchy Process). In this study, for each considered technology, several criteria were given weights, amongst which the pollutant emissions criterion. Thus, the lower the weight, the higher the amount of pollutants emitted. Table 9 presents the relevance weights of the criteria of pollutant emission, in the scale of Marchezetti's work, for the technologies studied in the present study, besides a general relation concerning the amount of pollutants emitted by each technology.

Table 9- Emission weights in the technology prioritization study

Source	Weight of the Emissions in the Ranking Study		
	Anaerobic Digestion	Incineration	Pyrolysis
Marchezetti, 2009.	0,081	0,040	0,087
General Relationship	(Anaerobic Digestion < Incineration) > Pyrolysis		

Source: Adapted from Marchezetti, 2009.

The weights data, presented in Table 9, serve as a way to establish a general relationship, also presented, of the technologies studied. To corroborate this relationship, the types of emissions of each technology can be observed

3.4.2 Types of Emissions

Marchezetti's (2009) study further presents, as an annex, a survey of the types of pollutants emitted by each technology. In this compilation one can observe the great diversity of incineration pollutants. In relation to other technologies, anaerobic digestion presents the emission of methane and acid gases, a fact that is more impactful to the environment than inert emissions such as ash and slag from pyrolysis technology. Table 10 presents the information gathered by Marchezetti, in addition to the relationship between the technologies in terms of the types of pollutants emitted.

Table 10- Types of pollutants emitted by MSW treatment technologies

Source	Pollutants Emitted		
	Anaerobic Digestion	Incineration	Pyrolysis
Marchezetti, 2009.	Produces CH_4 e H_2S and other gases.	$CO_2, SO_x, N_2,$ CH_4, H_2S dioxins, furans, ashes, unburnt residues and other gases.	Ash and slag.
General Relationship	(Anaerobic Digestion < Incineration) > Pyrolysis		

Source: Adapted from Marchezetti, 2009.

3.4.3 Ranking based on emissions

The partial ranking based on emissions was performed by means of the relations defined both by the weights in the multicriteria analysis and by the types of emitted pollutants. The relations in both criteria were equal, being also the general relation of the parameter "emissions". Technologies were classified from the lowest to the highest quantity and diversity of emitted pollutants. Table 11 shows both the relations obtained in each criterion and the general relation, as well as the positioning and score in the partial ranking based on emissions.

Table 11- Partial ranking of the technologies based on the "Emissions" parameter

Parameter: Emissions	Partial Ranking		
	Anaerobic Digestion	**Incineration**	**Pyrolysis**
Weights in Multicriteria Analysis	(Anaerobic Digestion < Incineration) > Pyrolysis		
Types of Pollutants Emitted	(Anaerobic Digestion < **Incineration**) > Pyrolysis		
General Relationship	(Anaerobic Digestion < Incineration) > Pyrolysis		
Position	2º	3º	1º
Score	**2 points**	**1 point**	**3 points**

Source: Author, 2021.

3.5 Logistics Analysis

In the analysis regarding logistics, several factors were considered, such as: (i) Installation Size, (ii) Operating Temperature, (iii) Operating Time, (iv) Volume Reduction, (v) Usual Applicability and (vi) Extra Activities. For each criterion, based on the arguments presented, a relationship between the technologies was stipulated. By means of these stipulated relations, at the end, it was possible to establish a general relation and a consequent partial ranking for the "Logistics" parameter

3.5.1 *Installation Size*

According Pedroxa (2017), one of the positive points of pyrolysis technology, as well as incineration, is the issue of small space for the installation of the enterprise. These plants can have small dimensions, and can be installed in relatively small sheds inside industrial areas.

Anaerobic digestion plants, on the other hand, due to their energy efficiency and the need to store a larger volume during decomposition, are feasibly outperformed by thermal technologies in terms of plant size.

Between incineration and pyrolysis, since the latter generally presents a larger quantity of equipment due to gas treatments and by-product control, it was ranked worst in terms of facility size, although both are close in size.

Thus, in as much as plant size is concerned, incineration, since it requires a smaller size than other technologies, is better classified. Next is pyrolysis, with dimensions relatively close to those of incineration. And finally, anaerobic digestion with greater space requirements (Incineration, Pyrolysis and Anaerobic Digestion).

3.5.2 Operating Temperature

According to Mota (2015), slow pyrolysis is characterized by small heating rates with temperature range around 600°C. Marchezetti (2009) adds that incineration operates in a range between 750°C and 1200°C and anaerobic digestion between 55°C and 60°C.

Therefore, as the lower the operating temperature, the lower is also the complexity of the technology in this criterion. Thus, the technologies were classified from the lowest to the highest operating temperature required. Thus, anaerobic digestion is the best classified, for having a lower need to raise the temperature, and is followed, in this order, by pyrolysis and incineration (Anaerobic Digestion, Pyrolysis and Incineration).

3.5.3 Operating Time

Slow pyrolysis (the pyrolysis process considered in this study) usually takes hours, while incineration usually takes minutes (DIP, 2004). The anaerobic digestion, in turn, due to the long period for decomposition of waste, usually has its process occurring over days.

Given that the shorter the operational time, the quicker the return on investment of technologies, these were ranked from the shortest to that of greatest operational time. Thus, incineration ranked highest since it usually requires only a few minutes to process a given quantity of SUW. Next comes pyrolysis and anaerobic digestion, which normally take hours and days, respectively, to process the same quantity of SUW (Incineration, Pyrolysis and Anaerobic Digestion).

3.5.4 Volume Reduction

According to the survey conducted by Marchezetti (2009), incineration has the ability to reduce up to 99% of the initial volume of waste and pyrolysis up to 90%, while anaerobic digestion reduces organic matter by up to 60% of its initial volume.

Given that the PNRS has as one of its objectives the reduction of the quantity of SUWs destined for sanitary landfills, the greater the volume reduction provided by the technology, the better this will be classified. Thus, incineration comes first, for reducing up to 99%, followed by pyrolysis, with reductions of up to 90% and lastly, anaerobic digestion, with maximum capacities of 60% reduction (Incineration, Pyrolysis and Anaerobic Digestion).

3.5.5 *Usual Applicability*

Anaerobic digestion plants are commonly used more often for the treatment of animal waste on farms and not for the treatment of MSW in urban areas. And this is justified for logistical reasons, because in rural environments, the operating plant is closer to its source of clean raw material, the organic waste, and is also closer to the destination of its by-product, the biofetilizer. It is worth remembering that both the raw material and the by-product are heavy materials and of low added value, which makes their transportation more costly (BAIN & COMPANY, 2012).

Also according to Bain & Company (2012), pyrolysis is still an incipient technology worldwide, with few plants in operation and these usually being on a small scale. Furthermore, there is a lot of uncertainty in cost estimates given the rare directing of this technology towards the treatment of SUWs.

Incineration has some relevance in the international context, especially in developed countries. It is a technology commonly used in densely populated areas and also in regions with little space for landfills or with legislation that hinders their deployment (BAIN & COMPANY, 2012).

As the greater the usual applicability, the more tested and valid is the technology, these were then classified from most commonly applied to least used. Thus, incineration is the best positioned since it is used in several large scale projects. Next is anaerobic digestion, which despite being used in small projects, is widely validated in the rural sector. And finally, pyrolysis, since it is still an incipient technology (Incineration, Anaerobic Digestion and Pyrolysis).

3.5.6 *Extra Activities*

In anaerobic digestion, besides the investment in the digestion plant itself, it is necessary to invest in a screening plant, which is responsible for separating and sending only the organic waste to the digestion plant. (BAIN & COMPANY, 2012). In relation to thermal technologies, one notices that in incineration, the very heat of the process is used in the process of generating electricity, while in pyrolysis there is first the formation of a fuel and then this fuel is used for the generation of electricity.

As the smaller the number of extra activities, generally, the less complex the technology becomes, these were classified from the lowest to the highest need for extra activities. Thus, the best positioned is Incineration as it has inherent to the process, the heat that is responsible for energy generation. Next is pyrolysis, in which the fuel must first be generated to then obtain the necessary heat. And finally, anaerobic digestion, which in addition to everything else, also requires a prior screening plant (Incineration, Pyrolysis and Anaerobic Digestion).

3.5.7 *Ranking based on logistics*

To perform the partial ranking of the "Logistics" parameter, first the overall relation of technologies in this parameter was determined, using for this the classifications performed in each criterion. Table 12 gathers the rankings of each criterion and presents the overall ratio obtained for the "Logistics" parameter.

Table 12- Relationship between criteria of the "Logistics" parameter

Criteria	Technologies		
	Best Ranking	**Intermediate Position**	**Worst Classified**
Installation Size	Incineration	Pyrolysis	D. Anaerobic
Operating Temperature	D. Anaerobic	Pyrolysis	Incineration
Operating Time	Incineration	Pyrolysis	D. Anaerobic
Volume Reduction	Incineration	Pyrolysis	D. Anaerobic
Usual Applicability	Incineration	D. Anaerobic	Pyrolysis
Extra Activity	Incineration	Pyrolysis	D. Anaerobic
General Relationship	**Incineration**	**Pyrolysis**	**Anaerobic Digestion**

Source: Author, 2021.

The general relation presented in Table 12 was obtained by means of the classifications in each criterion. That is, as in most criteria, incineration was the best ranked, therefore this technology also occupies the best ranking in the overall relation. The same criterion was used for the intermediate position and for the worst ranked technology.

Based on the general relation of Table 12, one could finally perform the partial ranking based on the "Logistics" parameter. Following the same methodology of the other parameters, i.e., the best ranked technology obtained 3 points, the one in intermediate position, 2 points and finally, the worst ranked one, 1 point. The partial ranking based on the "Logistics" parameter is presented in Table 13.

Table 13- Partial ranking of technologies based on the "Logistics" parameter

Parameter: Logistics	Partial Ranking		
	Anaerobic Digestion	Incineration	Pyrolysis
General Relationship	Anaerobic Digestion < (Incineration > Pyrolysis)		
Position	3º	1º	2º
Score	1 point	3 points	2 points

Source: Author, 2021.

3.6 Social Analysis

According to Leite (2016), whilst anaerobic digestion works in conjunction with selective collection, thermal technologies compete for the same material valued by recycling, since they depend on such waste for greater efficiency. Between incineration and pyrolysis, given the fact that incineration has its revenue exclusively related to the calorific power of waste, this competition with recycling is more sensitive.

With respect to labor, the same line is followed. Anaerobic digestion has a larger number of employees, especially considering the recycled material collectors. And among thermal technologies, both require a more qualified labour force, pyrolysis, for having more activities in the process, being a greater recruiter.

Thus, technologies were ranked from the most to the least affinity with selective collection and inclusion of workers. Table 14 presents the relation between technologies as well as their classification and score in the "Social" parameter.

Table 14- Partial ranking of technologies based on the "Social" parameter

Parameter: Social	Partial Ranking		
	Anaerobic Digestion	Incineration	Pyrolysis
Milk, 2016.	Anaerobic Digestion > (Incineration < Pyrolysis)		
Position	1º	3º	2º
Score	3 points	1 point	2 points

Source: Author, 2021.

3.7 Overall Ranking and Technology Selection

After performing the partial rankings, it is possible, by summing the scores obtained in each parameter, to determine the overall score of the technologies. Table 15 presents the scores obtained in the analyzed parameters, in addition to the total score.

Table 15- Scores of each SUW treatment technology

Criteria	Score		
	Anaerobic Digestion	Incineration	Pyrolysis
Costs	3	1	2
Recipes	1	2	3
Emissions	2	1	3
Logistics	1	3	2
Social	3	1	2
Total Score	10 points	8 points	12 points

Source: Author, 2021.

Given the total score, it was finally possible to determine the overall ratio of technologies and the final ranking, thus selecting the technology used in the case study of this work. Table 16 shows the final ranking of the technologies.

Table 16- Final ranking of the MSW treatment technologies

Final Ranking			
Technology	Anaerobic Digestion	Incineration	Pyrolysis
Score Considered	10 points	8 points	12 points
General Relationship	(Anaerobic Digestion > Incineration) < Pyrolysis		
Position	2º	3º	1º

Source: Author, 2021.

Given this detailed selection, the Pyrolysis technology, more specifically the slow pyrolysis process, is the MSW treatment with energy recovery applied in this paper's case study.

Estimates, evaluations and conclusions concerning the implementation of MSW treatment at SISOL in Amazonas, are based on this technology of decomposing matter, at high temperatures, in the absence of oxygen and seeking the transformation of MSW into a solid fuel used to generate electricity.

4 CHARACTERIZATION OF THE AMAZON SISOL

Once the technology considered in the case study has been defined, the characterization of SISOL is required, addressing several variables that are necessary to determine the feasibility, or not, of implementing an alternative source of energy from MSW.

Recognizing that, as presented by Crispim (2019), there are numerous challenges to the collection of SUW data and information at Brazilian municipalities, since such data is often non-existent or precarious, it is reasonably understandable that for SISOL, an even greater characterization difficulty is observed.

Therefore, the purpose of this chapter is to present which information is required to analyse the implementation of the pyrolysis technology in SISOL, from where it was collected or how it was estimated.

4.1 SUW generation *per capita*

When one wishes to determine the quantity of electrical energy that will be generated as of SUW processing, it is intuitive to imagine that one necessary piece of information is the quantity of SUW available for the operation, usually this data is informed as SUW *per capita* generation.

This is a very difficult information to find regarding the municipalities of the state of Amazonas. Logically, this data is even rarer when it comes to SISOL. One of the few official information found in the literature was a study presented by the National Sanitation Information System - SNIS.

This study, called "Diagnosis of the management of Urban Solid Waste", in 2018, presented the value of MSW generation for 28 municipalities of the state of Amazonas. These data are organized in the table of ANNEX B.

Logically, the information from the SNIS alone is not enough to characterize all the SISOLs. However, these data are crucial to be used as a means of validating SUW generation estimate models.

A variety of models for estimating the *per capita* generation of SUW in a given locality can be found in the literature. Table 17 presents and organizes some of these models that have been used in recent years.

Table 17- Models for estimating SUW generation *per capita*

Models	Formulation
MARTINEZ et al., 2006.	$RSU_{pc} = 0{,}00006 . PIB + 0{,}5656$
MELO et al. 2009.	$RSU = (1.5657 . PIB - 3{,}6861 . POP + 5{,}5416) * {}^{1}\!/_{POP}$
DIAS et al., 2012.	$RSU_{pc} = -0{,}00000005 . RPC^2 + 0{,}0006 . RPC + 0{,}2848$
SOARES et al., 2015.	$RSU_{pc} = 0{,}00484 . POP + 0{,}1208 . PIB + 2{,}716 . IDHM - 1{,}983$
JUNIOR et al.,2018.	$RSU_{pc} = 6{,}19 . 10^{-9} . POP + 0{,}0162 . RPC + 0{,}00422 . C_e + 0{,}339$

Source: Adapted from MARTINEZ et al. (2006), MELO et al. (2009), DIAS et al. (2012), SOARES et al. (2015) and JUNIOR et al. (2018).

Table 17 demonstrates that *per capita* SUW generation models are a function of several variables such as: population, *per capita* income, electricity consumption etc. Thus, to utilize models, one must first determine these variables, whether by direct survey (whenever possible) or by estimates (which is the majority of cases).

In addition, many of these variables are necessary data for the analyses of this work. Therefore, explaining how this information was obtained or estimated is precisely the function of this characterization chapter. Thus, in the following, it is explained how the data were collected: (i) population, (ii) GDP, (iii) income *per capita*, (iv) municipal HDI and (v) electricity consumption *per capita*.

(i)   Population

The estimate of the population of SISOL, was given, initially, determining the population of the municipalities of the Amazon. In 2010, the Brazilian Institute of Geography and Statistics - IBGE, through the Demographic Census, presented the numbers of each municipality.

Over the years, the IBGE presented the population estimates of the municipalities, more precisely estimates from 2011 to 2019 were presented. As the present work aims to perform analyses until 2024, it was necessary to make a projection of the population for the remaining years.

38

This projection was performed by forecast calculations of future values, based on a history of values (the IBGE data from 2010 to 2019). For these calculations, an exponential smoothing algorithm (ETS) is used. In a practical way, the projection was carried out using the Microsoft Excel function "PREVIEW.ETS". Both the IBGE data and the projection of the other years are gathered in the first table of APPENDIX H.

IBGE also presents a projection of the total population of the state of Amazonas. This projection, starting in 2011, encompasses all the years covered in this work. In this way, it was possible to use this IBGE projection to validate the projection of municipal populations performed through the ETS algorithm.

For this validation, the values of the IBGE projection (2011-2019) were compared with the values of IBGE's own estimate (2011-2019). And, logically, the data from the IBGE projection (2020-2024) were also compared with the values of the projection with the ETS algorithm (2020-2024).

In the first case, in which data from IBGE itself is compared, an error of approximately 1.32% was observed. In the comparison between the IBGE projection and the one performed in this work, the observed error was 1.52%. Both errors, smaller than 2.00%, are acceptable, and validate the projection with the ETS algorithm.

Some municipalities are SISOL, but the other SISOL are districts or villages belonging to some municipality. The estimation of the population of the districts and villages was done through a proportion using electricity consumption as a basis.

In the table in ANNEX A, it is possible to see which municipalities belong to the SISOL districts and villages. In addition, it is possible to observe the energy demand of SISOL, both the municipalities and the districts and villages.

The   energy demand and the estimated population of the municipalities were used to calculate the *per capita* consumption in a given municipality. Given this *per capita* consumption and the energy demand of the district or town (ANNEX A) it is possible to finally determine the population of SISOL that are districts and towns.

It is worth remembering that it was necessary to subtract the population of the districts and villages from the total population of their respective municipalities to avoid the problem of double counting. With this, the populations of SISOL were estimated. The values are also organized in APPENDIX H.

(ii)   GDP

Regarding the Gross Domestic Product - GDP, the IGBE provided estimates from 2010 to 2018. Therefore, for the remaining years (2019 to 2024) a projection was also performed using the ETS algorithm.

Using the estimated population information, it was possible to determine the GDP *per capita*. Based on the *per capita* GDPs of the municipalities, and the SISOL populations, it was possible to estimate the total GDP of all SISOLs.

The table in APPENDIX I shows the GDPs, both in dollars and reais, of the SISOLs in the Amazon. In addition, the table also presents the GDP *per capita of* each location.

(iii)   *Per capita* income

As for *per capita* income - RPC, the only precise information found was that presented by IBGE in the 2010 Census. With only this information, it is not possible to make a projection through the ETS algorithm, since this projection is a prediction of future value based on a history of values.

To overcome this problem of lack of historical values, the data of PRC of the state of Amazonas was used. These values represent the average PRC of all municipalities of the state. These data were presented from 2014 to 2019, by IBGE, through the National Continuous Household Sample Survey - PNADc.

To determine the municipal PRC based on the state one used the proportion system based on the 2010 Census data. This proportion is given by the use of a factor value, which relates the municipality's PRC in 2010 to the state's average PRC in 2010.

For example, the municipality of Parintins in the 2010 Census had a PRC of R$400.86, while the state PRC was R$327.41. The factor value of Parintins is given by the ratio between these numbers, therefore, 1.224. To estimate the municipal PRC in the coming years, the product between the factor value of the municipality and the state PRC of the year in question (provided by the PNADc) was performed. Therefore, the PRC of Parintins was estimated at R$904.78 in 2014 (state PRC R$739.00), R$921.92 in 2015 (state PRC R$753.00) and so on.

Thus, the RPCs of the municipalities from 2014 to 2019 were estimated. Thus, the historical values necessary to perform the projection with the ETS algorithm were obtained. This projection was performed for the years 2020 to 2024. The RPCs obtained were replicated for the districts and towns according to the municipalities to which they belong.

In APPENDIX J are organized: the 2010 Census PCRs, the state PCRs provided by PNADc, the factor values of each locality, besides, of course, the estimated and projected PCRs for SISOL.

(iv)   HDI

The Municipal Human Development Index - MFDI, is a data that is usually presented in the Demographic Census, i.e., usually every 10 years. It is information derived from several others such as: life expectancy, education and GDP.

Since it is an index, this data varies on a scale of 0 to 1, which makes any kind of trivial projection impossible. However, the HDI is a data that presents a relatively small change over the years, thus, the literature usually uses in its analysis, the last official update of the index.

The last HDIM update was presented by the IBGE, through the 2010 Census. For the districts and villages, as in previous cases, the HDMI of the municipality to which they belong was considered. In the table in APPENDIX K, the HDMI of the SISOL Amazon are organized.

(v)   **Electrical Consumption** *per capita*

As already addressed in the population item, the *per capita* electricity consumption was estimated using the information of energy demands of SISOL that are organized in ANNEX A, along with the population data estimated in this work. In APPENDIX L, the *per capita* electric consumption values for SISOL are organized.

Once these variables were estimated, it was finally possible to use the models in Table 17. Through each model, the values of SUW generated per inhabitant per day were estimated for the 28 municipalities that are included in the 2018 SNIS study. The results of SUW generation *per capita* for each model are organized in the table in APPENDIX M

By comparing the results of each model (APPENDIX M) with the actual values presented by SNIS (APPENDIX B) it was possible to measure the average variations. The lower the average value of the model variation, the closer the estimates of the model in question are to reality. Table 18 presents the average variations of each model analyzed.

Table 18- Average variations of model results when compared to SNIS

Models	Average of the variations
Model 01 - MARTINEZ et al., 2006.	0,597
Model 02 - MELO et al. 2009.	2211972,482
Model 03 - DIAS et al., 2012.	0,537
Model 04 - SOARES et al., 2015.	7102688,250
Model 05 - JUNIOR et al.,2018.	1,747

Source: Author, 2021.

Table 18 notes that two of the models (2 and 4) presented absurd variation results because their estimates of SUW *per capita* gave values that were totally outside the reality of any locale. The remaining models on the other hand, resulted in acceptable values and Model 02 was the one that presented the lowest average variation value.

Given this, the model selected to estimate *per capita* SUW generation at SISOL was Model 03, presented by Dias in 2012, and which uses CRP as its analysis variable. Thus, *per capita SUW* generation of SISOL in the Amazon was estimated and presented in APPENDIX N.

4.2 Gravimetric Composition

Once the quantity of SUW generated by each SISOL is estimated, another necessary information that is quite intuitive is the gravimetric composition of this SUW. The gravimetric composition, in simpler words, represents the percentage of each component in a sample of waste.

Logically, if the quantity of MSW from Amazonian SISOL is already a non-existent information in literature, the gravimetric composition of MSW is also not found in a straightforward manner.

In fact, many studies are faced with the difficulty of obtaining this type of information. What is usually done is to use an average gravimetric composition that has been previously studied and validated.

From this perspective, a global study that presents a validated average gravimetric composition is that presented by Hoornweg and Bhad-Tata in the book *"What a waste"*. In this study, the authors attributed an average gravimetric composition for each income level of the population in question. Table 19 presents these averages according to local income.

Table 19- SUW Gravimetric Composition according to Local Income

	Low Income	Medium-Low	Medium-High	High Income
Organic	64,0%	59,0%	54,0%	28,0%
Paper	05,0%	09,0%	14,0%	31,0%
Plastic	08,0%	12,0%	11,0%	11,0%
Glass	03,0%	03,0%	05,0%	07,0%
Metal	03,0%	02,0%	03,0%	06,0%
Other	17,0%	15,0%	13,0%	17,0%

Source: HOORNWEG; BHADA-TATA, 2012.

Table 19 shows that the lower the local income, the higher the percentage of organic matter in the waste. On the other hand, the higher the income, the greater are the shares of paper and plastic. Generally these latter materials come from packaging and other products discarded by the class of greater consumer power.

To be able to use this type of classification in the study of SISOLs, it is necessary to classify them in some kind of local income class. To do so, the RPC data estimated in APPENDIX J were used. The classification was made in a similar way to what happened before 2014, when social classes were organized into high, middle, low, etc.

Therefore, if the SISOL PRC is lower than the minimum wage in the year in question, this locality is classified as Medium Low or Low Income. To be classified as Low Income, the PRC must also be less than $\frac{1}{4}$ of the minimum wage in the analyzed period.

If the RPC of SISOL is higher than the minimum wage in force, the location in question is classified as Medium High or High Income. To be classified as High Income, the PRC must exceed four times the minimum wage.

In this way, each SISOL was classified for each specific year. These classifications, for the years relevant to this study, are organized in APPENDIX O.

As a way to validate this type of classification, a study by the Government of Amazonas presented in the Solid Waste and Selective Collection Plan of the Metropolitan Region of Manaus (PRSCS-RMM), in 2017, was used.

In this study is informed the average gravimetric composition of the Metropolitan Region of Manaus - RMM. Table 20 presents the gravimetric composition of the MRM, presented in the study of the Government of Amazonas.

Table 20- Average Gravimetric Composition of RMM

Component	RMM
Organic	47,3%
Paper	13,3%
Plastic	14,1%
Glass	01,3%
Metal	03,2%
Other	20,8%

Source: PRSCS-RMM, 2017.

Comparing it with the gravimetric compositions of Hoornweg and Bhad-tata (Table 19), and averaging the variations, a variation of 6.1% was obtained in the Low Income class; 4.5% in the Lower Middle Income class; 3.7% in the Upper Middle Income class and 8.7% in the High Income class.

Thus, the RMM is classified according to the class that presented the lowest average variation, thus it was classified as Medium High. To complete the validation, the RPC of the MMR was then analyzed.

The MMR consists of 13 Amazonian municipalities. These municipalities, as well as the PRC of each of them in the year 2017, are organized in Table 21.

Table 21- *Per capita* income of the Municipalities that make up the MMR

RMM	RPC
Autazes	R$ 762,63
Careiro	R$ 650,09

Careiro da Várzea	R$ 664,19
Iranduba	R$ 1.078,66
Itacoatiara	R$ 1.236,81
Itapiranga	R$ 899,14
Manacapuru	R$ 1.086,14
Manaquiri	R$ 847,63
Manaus	R$ 2.369,29
New Airão	R$ 821,12
President Figueiredo	R$ 1.325,11
Rio Preto da Eva	R$ 988,29
Silves	R$ 825,80

Source: Author, 2021.

The MMR's PRC is precisely the average of the PRCs of its 13 municipalities, so, in 2017, this value was R$1,042.68. The minimum wage in 2017 was R$937.00, i.e., the RPC of the MMR is greater than the minimum wage at the time and less than four times its value. Thus, the MMR is also classified as Medium High, thus validating the technique used to classify the SISOL in APPENDIX O.

Given all that has been exposed, the objectives of this chapter in characterizing the SISOL are concluded, raising data and information necessary for the analysis and conclusions of subsequent chapters.

5   RESULTS

Once the pyrolysis technology was selected and the information about SISOL was collected, it was finally possible to perform the economic analysis in the certainty condition. For this, all the investments, revenues and taxes were related in a cash flow, being able to determine the indicators of economic viability of the project.

5.1 Investments

The investments for the project include both the expenditures with the implementation of the enterprise, as well as the operational expenditures. In addition, depreciation of equipment and buildings should also be considered.

5.1.1   Implementation costs

Using the proportion of values according to the plant capacity, adapted from Lameu(2018), together with the plant value commercialized by the Italian company *Maim Engineering*, we estimated, in reals, the values of the pyrolysis plants according to their capacities. These values are organized in Table 22.

Table 22- Prices of pyrolysis plants according to capacity

Capacity (kg/h)	Value (US$)	Value (R$)
40	454.552,45	2.431.855,60
200	909.104,90	4.863.711,19
500	1.515.174,83	8.106.185,32
1000	2.424.279,72	12.969.896,52
2000	3.939.454,55	21.076.081,84
2500	4.545.524,48	24.318.555,97

Source: Adapted from Lameu (2018).

Considering Consema Resolution 13/2012, this type of enterprise is classified as small-sized and of great polluting potential, which entails the need, for its licensing, of an Environmental Impact Study and Environmental Impact Report - EIA/RIMA.

According to CCaibre (2016), the cost of the environmental study and other documents, as well as the analysis and inspection fees of the state environmental agency, represents about 3% of the value of the pyrolysis plant in the supplier company.

Taking into account the same study, the execution of the civil construction, including the impermeable area for the pyrolysis plant installation, sheds, office areas, among others, has a cost around 15% of the plant's value.

Given this, Table 23 presents the expenses for the implementation of the undertaking of energy recovery from MSW, by means of pyrolysis.

Table 23- Summary of the expenses with the implementation of the project

Investments	Value
Pyrolysis unit (plant)	Table 22 Value
Engineering projects and environmental permits	3% of the plant value
Civil Construction	15% of the plant value

Source: Author, 2021.

5.1.2 Operating Expenses

According to CCarta (2012), the cost of inputs and administrative expenses, which are indicated as variable operating costs in the venture's cash flow, represents about 6.5% of the capital invested in the deployment.

Also according to the study, the maintenance cost, which is designated in the cash flow as fixed operating cost, is approximately 4.5% of the capital invested in the implementation of the enterprise.

Finally, the cost of labor, takes into consideration the salaries of employees, as well as social and labor charges. Social charges include contributions such as INSS (National Institute of Social Security), FGTS (Guarantee Fund for Time of Service), among others.

Labor charges, on the other hand, involve: 13th salary, vacations, remunerated weekly rest period, licenses, aids, work accidents, legally abonished absences, prior notice, etc.

According to Caibre (2016), wages along with social and labor charges represent approximately the same fixed operating cost of maintenance, i.e., 4.5% of the capital invested in the implementation of the project.

Given this, Table 24 presents the summary of operational expenses of the undertaking of energy recovery from MSW, by means of pyrolysis.

Table 24- Summary of operating expenses of the enterprise

Investments	Annual Value
Inputs and Other Expenses	6.5% of the deployment expenditure
Maintenance	4.5% of the deployment expenditure
Labor	4.5% of the deployment expenditure

Source: Author, 2021.

5.1.3 Depreciation

Considering SRC Normative Instruction 162 of December 31, 1998 of the Federal Revenue Service, which discusses the useful life of each type of asset, and analyzing the characteristics of the equipment of the development, a useful life of 10 years was determined for the equipment and 25 years for the buildings.

Given the amounts spent on equipment and buildings, along with their useful lives, it was possible to determine the annual value of the depreciation of each item. These values are organized in Table 25.

Table 25- Summary of depreciation expenses

Item	Implementation Expenses	Useful Life	Annual Depreciation
Equipment	Table 22 Value	10 years	10% of the value of Table 22
Buildings	3% of the plant value	25 years	0.12% of the plant value

Source: Author, 2021.

5.2 Recipes

The revenues of the undertaking of energy recovery from MSW, by means of pyrolysis, considered in this study, are those obtained through the sale of generated electric power and the sale of the biofertilizing by-product.

5.2.1 *Sale of electricity*

It was considered as the sale price, the average of the energy sale values, in the northern region, traded in the electronic energy sales exchange of the CCEE - Electrical Energy Trading Chamber, in the last 12 months (second semester of 2020 and first semester of 2021).

In this period, the price of energy varied between R$55.57/MWh and R$502.70/MWh. This large oscillation occurred due to the impact on trading caused by the Covid-19 pandemic. In the present work was used then, the average price of the period which was R$199.52/MWh.

Regarding the pyrolysis plant generation efficiency, according to Caibre (2016), it is possible to generate 01 MW of energy for each ton of MSW (considering organic matter) processed.

5.2.2 *Sale of Biofertilizer*

Also in the study by Caibre (2016), the biofertilizer yield factor per ton of MSW (considering organic matter) was determined, with this factor being equal to 12%. That is, approximately 120kg of biofertilizers are produced for each ton of MSW processed. In relation to the sale price, the study estimated an average value of R$125.00 per ton of fertilizer.

Thus, the revenues of the project, both for the sale of electricity and the sale of biofertilizer, are schematized in Table 26.

Table 26- Summary of revenues of the enterprise

Recipe Type	Efficiency	Price
Sale of Energy	1t of MSW = 1MW	R$199.52/MWh
Sale of Biofertilizer	1t of MSW = 120kg	R$125.00/t

Source: Author, 2021.

5.3 Taxes

There are several taxes that have rates levied on some of the operations of the pyrolysis plant, both related to the generation and sale of electric power and the production and sale of biofetilizer. Some of these rates are levied on only one of the operations, others on both.

Examples of these taxes are: COFINS - Contribution for Social Security Financing; CSLL - Social Contribution on Net Profit; ICMS - Value-Added Tax on Sales and Services; IPI - Tax on Industrialized Products; IR - Income Tax and PIS/PASEP - Social Integration and Public Service Employee Savings Program.

Table 27 presents the tax rates levied on each type of operation of the enterprise.

Table 27- Tax rates levied on project operations

Taxes	Electric Power	Biofertilizer
COFINS	7,60%	0,00%
CSLL	9,00%	9,00%
ICMS	25,00%	12,00%
IPI	0,00%	Not Taxable
PIS/PASEP	1,65%	0,00%
IR	15% + 10% additional above R$20,000.00/month	

Source: Author, 2020.

5.4 Economic Analysis

For the economic analysis it was determined the cash flow of the enterprise, from 2019 to 2024. The year 2019 was considered the year of implementation of the project. In 2024, a sale of the enterprise was arbitrated, by the residual value, taking into account the depreciation in the period considered, without inflation correction.

Basically the cash flow follows the following schematization: first the annual Gross Revenue (RB) is determined, then the taxes on product sales (IPI, ICMS, PIS and COFINS) are subtracted obtaining the Annual Net Revenue (RLA).

The operating costs and depreciation are subtracted from the RLA, obtaining the so-called Taxable Net Profit (LTV). From the latter, income tax and social contribution are deducted, thus determining the Annual Real Profit (AKI) of the enterprise.

Having these annual cash flows and the expenses with the implementation of the venture, the Net Present Value (NPV) was determined, which represents the values of annual cash flows corrected for year zero, i.e., for the year of implementation of the venture (2019). This data is easily calculated by *Microsoft Excel*, through the "NPV" function.

Another information that was determined in cases of positive NPV, was the Internal Rate of Return (IRR), this rate refers to the percentage of return on capital invested in the implementation of the venture. This information can also be determined through *Microsoft Excel,* by means of the function "IRR".

Thus, the NPV allows for the evaluation of whether a given venture is economically feasible, given a Minimum Rate of Attractiveness (TMA). In this study, it was considered a TMA of 6%, which is a value generally used as fixed income yields.

For a project to be considered economically feasible, it is necessary that the calculated NPV is positive, because its IRR will be greater than its TMA. In cases in which the IRR is less than the TMA, the calculated NPV will be negative, indicating the economic infeasibility of the project.

Thus, a financial flow was built for each of the SISOLs, aiming to determine the NPV of the enterprise in that location and consequently the project's economic feasibility. Figure 6 presents, as an example, the financial flow assembled for Parintins, the SISOL with the highest daily MSW generation.

Figure 6- Financial flow of SISOL in Parintins

PARINTINS

Descrição	2019	2020	2021	2022	2023	2024
Receita Bruta	-	R$ 3.859.205,21	R$ 4.084.257,83	R$ 4.091.141,92	R$ 4.337.022,93	R$ 4.343.627,28
(-)Impostos sobre Vendas	-	R$ 1.261.736,31	R$ 1.335.315,47	R$ 1.337.566,16	R$ 1.417.954,99	R$ 1.420.114,23
(=) Receita Liquda Anual	-	R$ 2.597.468,90	R$ 2.748.942,37	R$ 2.753.575,75	R$ 2.919.067,93	R$ 2.923.513,04
(-)Insumos e Outras Despesas	-	R$ 3.233.070,95	R$ 3.233.070,95	R$ 3.233.070,95	R$ 3.233.070,95	R$ 3.233.070,95
(-)Gastos com Manutenção	-	R$ 2.238.279,89	R$ 2.238.279,89	R$ 2.238.279,89	R$ 2.238.279,89	R$ 2.238.279,89
(-)Gastos com Mão-de-obra	-	R$ 2.238.279,89	R$ 2.238.279,89	R$ 2.238.279,89	R$ 2.238.279,89	R$ 2.238.279,89
(-)Depreciação Equipamento	-	R$ 4.215.216,37	R$ 4.215.216,37	R$ 4.215.216,37	R$ 4.215.216,37	R$ 4.215.216,37
(-)Depreciação Edificações	-	R$ 50.582,60	R$ 50.582,60	R$ 50.582,60	R$ 50.582,60	R$ 50.582,60
(+)Venda de Ativos	-	-	-	-	-	R$ 28.410.558,32
(=)Lucro Liquido Tributável	-	-R$ 9.377.960,80	-R$ 9.226.487,34	-R$ 9.221.853,95	-R$ 9.056.361,77	R$ 19.358.641,66
(-)IR	-	0	0	0	0	R$ 4.839.660,42
(-)CSLL	-	0	0	0	0	R$ 1.742.277,75
(=)Lucro Real	-	-R$ 9.377.960,80	-R$ 9.226.487,34	-R$ 9.221.853,95	-R$ 9.056.361,77	R$ 12.776.703,50
INVESTIMENTOS	R$ 49.739.553,15	-	-	-	-	-
(-)Unidade de Pirólise	R$ 42.152.163,69	-	-	-	-	-
(-)Licenças Ambientais	R$ 1.264.564,91	-	-	-	-	-
(-)Construção Civil	R$ 6.322.824,55	-	-	-	-	-
Fluxo de Caixa	-R$ 49.739.553,15	-R$ 9.377.960,80	-R$ 9.226.487,34	-R$ 9.221.853,95	-R$ 9.056.361,77	R$ 12.776.703,50

V.P.L.	-R$ 72.167.063,92	T.I.R.	-	US$1,00 = R$5,35

Source: Author, 2021.

Figure 6 shows that generating electricity from MSW pyrolysis at SISOL in Parintins is economically unfeasible, since it has a negative NPV (-R$72,167,063.92), according to the annual cash flows presented.

51

Figure 7 presents the financial flow of the project to generate electric power from MSW, for the SISOL of Vila Bitencourt, a location that presented the lowest daily generation of MSW.

Figure 7- Financial flow of SISOL of Vila Bitencourt

VILA BITENCOURT

Descrição	2019	2020	2021	2022	2023	2024
Receita Bruta	-	R$ 7.061,92	R$ 6.790,22	R$ 6.969,62	R$ 7.038,96	R$ 7.040,41
(-)Impostos sobre Vendas	-	R$ 2.308,84	R$ 2.220,01	R$ 2.278,66	R$ 2.301,33	R$ 2.301,81
(=) Receita Líquida Anual	-	R$ 4.753,08	R$ 4.570,21	R$ 4.690,96	R$ 4.737,63	R$ 4.738,60
(-)Insumos e Outras Despesas	-	R$ 186.523,32	R$ 186.523,32	R$ 186.523,32	R$ 186.523,32	R$ 186.523,32
(-)Gastos com Manutenção	-	R$ 129.131,53	R$ 129.131,53	R$ 129.131,53	R$ 129.131,53	R$ 129.131,53
(-)Gastos com Mão-de-obra	-	R$ 129.131,53	R$ 129.131,53	R$ 129.131,53	R$ 129.131,53	R$ 129.131,53
(-)Depreciação Equipamento	-	R$ 243.185,56	R$ 243.185,56	R$ 243.185,56	R$ 243.185,56	R$ 243.185,56
(-)Depreciação Edificações	-	R$ 2.918,23	R$ 2.918,23	R$ 2.918,23	R$ 2.918,23	R$ 2.918,23
(+)Venda de Ativos	-	-	-	-	-	R$ 1.639.070,67
(=)Lucro Líquido Tributável	-	-R$ 440.033,31	-R$ 440.216,18	-R$ 440.095,43	-R$ 440.048,76	-R$ 440.047,78
(-)IR	-	R$ 0,00	R$ 0,00	R$ 0,00	R$ 0,00	R$ 0,00
(-)CSLL	-	R$ 0,00	R$ 0,00	R$ 0,00	R$ 0,00	R$ 0,00
(=)Lucro Real	-	-R$ 440.033,31	-R$ 440.216,18	-R$ 440.095,43	-R$ 440.048,76	-R$ 440.047,78
INVESTIMENTOS	R$ 2.869.589,60	-	-	-	-	-
(-)Unidade de Pirólise	R$ 2.431.855,60	-	-	-	-	-
(-)Licenças Ambientais	R$ 72.955,67	-	-	-	-	-
(-)Construção Civil	R$ 364.778,34	-	-	-	-	-
Fluxo de Caixa	-R$ 2.869.589,60	-R$ 440.033,31	-R$ 440.216,18	-R$ 440.095,43	-R$ 440.048,76	-R$ 440.047,78

V.P.L.	-R$ 4.723.407,95	T.I.R.	-	US$1,00 = R$5,35

Source: Author, 2021.

Figure 7 shows that the implementation of the project of energy recovery from the pyrolysis of USW in SISOL of Vila Bitencourt has also shown to be unfeasible, since the negative NPV of -R$4,723,407.95.

The NPVs of SISOL in the Amazon are organized in APPENDIX P in the category "Current Scenario". It is noted that under the current conditions considered in this work, **NONE** of the SISOL in the Amazon presented a positive NPV for the project.

However, it is worth noting that some of these current conditions have suffered the impacts of the Covid-19 pandemic, such as the value of the dollar and the sale price of electricity. And some others can be adjusted by political powers, such as taxes.

Thus, it is relevantly important to perform a sensitivity analysis considering these adjustable parameters and re-analyzing the feasibility of the project.

5.5 Sensitivity Analysis

The sensitivity analysis aims to assess the economic impact of the variation of certain parameters that were used in the economic analysis in the condition of certainty.

In this work, the focus of the economic analysis was on the parameters that were heavily impacted by the Covid-19 pandemic or that can be adjusted by public authorities.

Thus, three scenarios were considered, in such a manner that at each scenario, one more parameter was adjusted, starting from the most easily altered to the most complex. Furthermore, the impacts of the variation of each parameter individually were observed.

5.5.1 *Scenario 01*

In the first scenario, the adjusted parameter was the sale price of electricity. This is because, during the pandemic, sales prices were frozen, which changed the average value used as a basis for calculation for the current scenario.

Thus, in Scenario 01, the highest sales price in the period considered was used, which was R$502.70/MWh, according to data presented by CCEE.

With the adjusted sale price of electricity, the financial flows of each SISOL were again determined. Table 28 presents the corrected NPVs for the SISOL of Parintins and Vila Bitencourt.

Table 28- NPVs of SISOL Parintins and Vila Bitencourt for Scenario 01

SISOL	Current Scenario	Scenario 01
Parintins	-R$72.167.063,92	-R$57.31.302,52
Vila Bitencourt	-R$4.723.407,95	-R$4.696.094,01

Source: Author, 2021.

Table 28 shows that despite the reduction in losses, the NPV remained negative in both example locations, indicating the economic unfeasibility of the project under Scenario 01 conditions.

It is also noticed that, as the SISOL of Parintins has a much higher daily MSW generation than Vila Bitencourt's SISOL, the variation of NPV with the change in the sale price of electricity was also much more sensitive.

The adjusted NPV values, for all SISOLs, are organized in APPENDIX P, in the category "Scenario 01". Similarly to the previous examples, **ALL NFSOLs** had their losses reduced, but remained with negative NPV.

5.5.2 *Scenario 02*

In the second scenario, the next parameter analyzed was the dollar value. This is because, the dollar price suffered a huge variation with the Covid-19 pandemic, it was observed an increase of over 29% only in 2020 (INVESTING, 2021).

In addition, the variation in the price of the dollar has a significant impact on implementation costs, since the values of the plants of the Italian company, *Maim Engineering,* are in dollars.

Thus, in Scenario 02, in addition to the variation in the sale price of electricity, a reduction of one third in the value of the dollar was arbitrated. Since this is a plausible reduction in view of the increase observed in the currency exchange rate during 2020 and 2021. Thus, the new quotation used was R$3.53.

The corrected NPV values according to the parameters of Scenario 02 are organized in Table 29.

Table 29- NPVs of SISOL Parintins and Vila Bitencourt for Scenario 02

SISOL	Current Scenario	Scenario 02
Parintins	-R$72.167.063,92	-R$16.118.461,60
Vila Bitencourt	-R$4.723.407,95	-R$2.353.750,19

Source: Author, 2021.

Table 29 shows that the reduction in the dollar rate managed to further reduce the loss, but not enough to make the project economically viable, since the NPVs remained negative.

This pattern was repeated, again, for **ALL** SISOL in Amazonas. The NPVs corrected according to the parameters of Scenario 02 are also organized in APPENDIX P, in the category "Scenario 02".

5.5.3 *Scenario 03*

In the third scenario, the impacts of taxes on the financial flows of each SISOL were evaluated. This is because, as already detailed in this paper, this type of enterprise has a high incidence of tax rates, which erode profits and increase losses.

As it is a project with a sustainable bias, it is plausible to consider, by the public authorities, the tax exemption on this type of enterprise. Working exactly as a government tax incentive for environmentally sustainable projects.

In this way, in Scenario 03, in addition to variations in the sale price of electricity and in the dollar exchange rate, the tax exemption for the project was also considered.

In Table 30, the corrected NPVs are organized according to the parameters of Scenario 03, for the examples Parintins and Vila Bitencourt.

Table 30- NPVs of SISOL Parintins and Vila Bitencourt for Scenario 03

SISOL	Current Scenario	Scenario 03
Parintins	-R$72.167.063,92	+R$864.061,60
Vila Bitencourt	-R$4.723.407,95	-R$2.206.594,79

Source: Author, 2021.

Table 30 shows that, according to the parameters of Scenario 03, the SISOL of Parintins presented a positive NPV, attesting to the feasibility of the project for this scenario, in addition to demonstrating the importance of tax incentives for environmentally sustainable projects.

Vila Bitencourt's SISOL, however, despite the slight reduction in losses, remained with a negative NPV, attesting to the fact that the venture's feasibility strongly influences the quantity of SUW generated by the locality in question.

The corrected NPV values according to the parameters of Scenario 03, for each of the Amazon SISO, are organized in APPENDIX P in the category "Scenario 03".

By analysing APPENDIX P one notices that 47 SISOL, that is, almost half (49.5%) of localities, presented a positive NPV, when adjusted according to the third scenario. These sites are precisely those with the largest amount of SUW generated on a daily basis.

It is worth remembering that this study considered a five-year activity horizon, and that when a longer period is considered, such as the lifetime of the buildings, forexample, an even greater number of locations may be observed where the project is economically feasible.

Another point to mention, is that were also analyzed, in addition to the scenarios previously mentioned, the individual variation of each of the adjusted parameters. However, no significant changes capable of altering the feasibility or otherwise of the project were observed.

6 CONCLUSION

The entire development of this work enabled several conclusions, in addition, of course, to the determination of the feasibility or not of generating electricity from MSW in SISOL Amazon. Some of these conclusions are as follows:

SISOL, in the context of electric energy, faces several problems such as: energy deficit, high energy prices and large pollutant emissions. These problems require resolution, even if not the energy recovery of SUWs by means of pyrolysis;

Another problem, not only of SISOL, but of most Amazonian municipalities, is the wrong destination of SUW, which goes against the objectives of PNRS, besides being a great waste of energy potential;

The selection of the technology performed in this research, based on parameters such as: costs, revenues, logistics, emissions and social parameters, determined the pyrolysis as the most appropriate technology for the project. This corroborates the idea that technologies that are still incipient are a fertile field of study for future work;

The existence or availability of data and information about the Amazonian municipalities and SISOL is rare and inefficient. Thus, the extensive work of data survey and estimation carried out in this research is an important source of information for future research on the most varied themes;

The alternative of generating electricity from MSW in SISOL did not prove to be economically feasible under current conditions. However, it is worth noting that some current parameters were greatly impacted by the Covid-19 pandemic, such as the sale price of electrical energy and the dollar exchange rate;

The sensitivity analysis conducted in this research also showed that tax incentive policies for environmentally sustainable projects are essential for the economic viability of many enterprises, including the project studied in this paper;

-The sensitivity analysis observed the change of three parameters (energy sale price, dollar exchange rate and tax exemption). In the scenario in which the three parameters were adjusted, economic viability was obtained in the implementation of the project in 49.5% of SISOL;

SISOL that managed to present an economic viability in the sensitivity analysis, were those with the highest daily generation of SUW. This occurs because the largest amount of SUW allows for the generation of greater income that is capable of meeting the high costs of the enterprise's implementation;

-Because it presents high implantation costs, an alternative that can improve the economic viability indexes, is to be able to implement the pyrolysis plant from the supply of national companies, which could present more accessible prices.

6.1 Suggestions for future work:

Several developments are possible from the logical sequence of the research of this work. Some of these developments that can be studied and analyzed by future works are:

Carry out economic analysis in the condition of certainty, as well as sensitivity analysis, using values, which may be ceded, from national companies;

-Assess projects that integrate the most different types of technologies for energy recovery from MSW;

-Systematize and organize, as well as update and correct, the data and information collected in the process of characterization of SISOL, so that they can serve as a source of secondary data for different researches;

To analyze possibilities of cost reductions, or increase in revenues, in the enterprises studied in this research, aiming to increase the indexes of economic viability;

-Investigate other possible solutions to the problems faced by SISOL in the energy field.

REFERENCES

ABREU, C. D.; HENKES, J. A. **Uma Análise Sobre O Tratamento De Resíduos Sólidos Urbanos: Proposta De Sistema Alternativo, Transformando Residuos Só-Lidos Em Carvão E Energia**. Revista Gestão & Sustentabilidade Ambiental, v. 8, n. 1, p. 1015, 2019.

AMARAL, F. M. **Biodigestão anaeróbia dos resíduos sólidos urbanos: um panorama tecnológico atual**. Instituo de Pesquisas Tecnológicas do Estado de São Paulo - IPT, 2004.

ARAÚJO, M.; SCHOR, T. **Resíduos de serviço de saúde no estado do amazonas: desafios para implantar sua gestão**. InterfacEHS - Revista de Saúde, Meio Ambiente e Sustentabilidade, v. 3, n. 1, p. 1-22, 2011.

BAIN & COMPANY, B. **Estudo Econômico-Financeiro para destinação final de Resíduos Sólidos Urbanos(RSU)**Belo HorizonteFundação Israel Pinheiro, , 2012.

CAIBRE, D. I. et al. **Analysis of economic feasibility of the pyrolysis process for treatment of municipal solid waste: a case study applied to a medium-sized city**. Revista de Ciências Ambientais, v. 10, n. 2, 16 dez. 2016.

CAIXETA, D. M. **Geração de energia elétrica a partir da incineração de lixo urbano: o caso de Campo Grande/MS**. Universidade de Brasília, 2005.

CALDERONI, S. **Os billions perdidos no lixo**. 4. ed. São Paulo: University of São Paulo, 2003.

CARTA, R.; CRUCCU, M.; SANNA, F. **Pirolisi lenta, humida e catalítica dela Pollina**, 2012.

CASTANHO, D. S.; ARRUDA, H. J. **Biodigestores rurais: modelos indiano, chinês e bateladaPonta** GrossaVI Semana de Tecnologia em Alimentos. Universidade Tecnológica Federal do Paraná, , 2002.

CCEE, C. DE C. DE E. E. **Report budget of sectorial accounts 2018 - CDE/RGR/CCC**São Paulo, 2017.

CNI, C. N. DA I. **Recuperação Energética De Resíduos Sólidos. Um Guia Para Tomadores**

De Decisão. Brasilia: .

COSTA, J. P. F. DA. **Mechanical and biological treatment of municipal solid waste: evaluation of its potential for the recovery of recyclable materials**. Universidade Nova de Lisboa, 2010.

CRISPIM, M. DO C. F. N. **Matrix of sustainability and analysis of environmental perception in relation to household solid waste in southwestern Amazonas**. Universidade Federal do Amazonas, 2019.

CUNHA, M. E. G. **Análise do setor de saneamento ambiental no aproveitamento energético de resíduos: o caso do município de Campinas**. UNICAMP, 2002.

DIAS, D. M. et al. **Modelo para estimativa da geração de resíduos sólidos domiciliares em centros urbanos a partir de variáveis socioeconômicas conjunturais**, 2012.

DINIZ, J. **Conversão térmica de casca de arroz à baixa temperatura: produção de bio-óleo e resíduo sílico-carbonoso adsorvente**. Universidade Federal de Santa Maria, 2005.

DIP, T. M. **Otimização de condições operacionais de processo visando a minimização da emissão de material particulado na incineração industrial de resíduos perigosos**. São Paulo: University of São Paulo, 6 oct. 2004.

EMPRESA DE PESQUISA ENERGÉTICA - EPE. **Avaliação preliminar do aproveitamento energético dos resíduos sólidos urbanos de Campo Grande, MSRio** de Janeiro, 2008.

EPE, E. DE P. E. **Planejamento do Atendimento aos Sistemas Isolados Horizonte 2024 - Ciclo 2019Rio** de JaneiroMinistério de Minas e Energia, , 2019. Available at: <http://www.epe.gov.br>

FERNANDO, A.; LIMA, S. DO C. Caracterização dos resíduos sólidos urbanos do município de Maxixe-Moçambique. **Caminhos de Geografia**, v. 42, n. jun, p. 335-345, 2012.

FRIGO, K. D. D. A. et al. **Biodigestores: Seus Modelos E Aplicações**. Acta Iguazu, v. 4, n. 1, p. 57-65, 2015.

FROTA, W. M. **Sistemas isolados de energia elétrica na Amazônia no novo contexto do**

setor elétrico brasileiro. State University of Campinas, 2004.

GRIPP, W. G. **Aspectos técnicos e ambientais da incineração de resíduos sólidos urbanos: considerações sobre a proposta para São Paulo**. University of São Paulo, 1998.

HENRIQUES, R. M. **Aproveitamento energético de resíduos sólidos urbanos: uma abordagem tecnológica**. Federal University of Rio de Janeiro, 2004.

HOORNWEG, D.; BHADA-TATA, P. **What a Waste: A Global Review of Solid Waste Management**. Knowledge Papers, 2012.

IDEC, I. B. D. D. C. **Consumo sustentável: manual de educaçãoBrasíliaConsumers** International, 2005.

INVESTING. **History of the dollar rate**. Available at: <https://br.investing.com/currencies/usd-brl-historical-data> Accessed on 07/17/2021.

IPT. **Lixo municipal: manual de gerenciamento integrado**. 2. ed. São Paulo: INSTITUTO DE PESQUISAS TECNOLÓGICAS, 2000.

LAJOLO, R. D. **Cooperativa de catadores de materiais recicláveis: guia para implantaçãoIPTSão** Paulo, 2003.

LAMEU, G. H. P. **Study on the implementation of a fast pyrolysis plant in a sugar and ethanol plant**. Toledo University Center, 2018.

LANÇA, R. O. **Diagnóstico e avaliação do potencial energético dos resíduos gerados em uma indústria alimentar**. Universidade Estadual Paulista, 2008.

LEITE, C. B. **Tratamento de resíduos sólidos urbanos com aproveitamento energético: avaliação econômica entre as tecnologias de digestão anaeróbia e incineração**. Universidade de São Paulo, 2016.

LIMA, B.; NEVES, G. **Raciocínio Lógico-Quantitativo e Matemática para Receita Federal**. 00. ed. Estratégia Concursos, 2019.

LIMA, L. M. Q. **Tratamento de lixo**. 2. ed. São Paulo: Hemus, 1991.

LOPEZ, A. A. **Estudo da gestão integrada de resíduos sólidos urbanos na bacia Tietê-Jacaré (UGRHI-13)**. Universidade de São Paulo, 2007.

MACHADO, C. F. **ncineração: uma análise do tratamento térmico dos resíduos sólidos urbanos de Bauru/Sp**. Universidade Federal do Rio de Janeiro, 2015.

MANCINI, P. J. P. **Uma Avaliação do sistema de coleta informal de resíduos sólidos recicláveis no município de São Carlos, SP**. São Paulo: University of São Paulo, 1999.

MARCHEZETTI, A. L. **Avaliação de alternativas tecnológicas para o tratamento de resíduos sólidos domiciliares pela aplicação do método AHP: Estudo de caso da região metropolitana de Curitiba**. Universidade Federal do Paraná, 2009.

MARTINEZ, J.; ADHIKARI, K. B.; BARRINGT, S. **Predicted growth of world food waste and methane productionNo Title,** 2006.

MAYER, B. **Characterization and anaerobic biodigestion of the organic fraction of solid waste generated in the Central Supply S.A. - Unit Foz do Iguaçu**. Federal Technological University of Parana, 2016.

MEDEIROS, V. A.; CASTRO, D. E. **Tecnologias de recuperação térmica e energética de resíduos sólidos.** Centro Federal de Ensino Tecnológico de Minas Gerais - CEFET-MG, , 2015.

MELO, L. A.; SAUTTER, K. D.; JANISSEK, P. R. **Estudo de cenários para o gerenciamento dos resíduos sólidos urbanos de Curitiba,** 2009.

MENEZES, R. A. A.; GERLACH, J. L.; MENDES, M. A. **Estágio Atual da Incineração no BrasilCuritibaVII** Seminário Nacional de Resíduos Sólidos e Limpeza Pública, , 2000.

MONTEIRO, J. H. P. et al. **Manual de Gerenciamento Integrado de Resíduos Sólidos.** 15. ed. Rio de Janeiro: .

MOTA, F. DE A. DA S. et al. **Pyrolysis of Lignocellulose Biomass: a Review.** Revista Geintec, v. 5, n. 4, p. 2511-2525, 2015.

MUTZ, D. et al. **Waste-to-Energy Options in Municipal Solid Waste Management. A guide for decision makers in emerging or developing countries.** Bonn: Deutsche Gesellschaft für

Internationale Zusammenarbeit (GIZ) GmbH, 2017.

ONS, O. N. DO S. E. **Annual plan of energy operation of isolated systems for 2020**Rio de Janeiro, 2019.

PAVAN, M. DE C. O. **Geração de energia a partir de resíduos sólidos urbanos: avaliação e diretrizes para tecnologias potencialmente aplicáveis no Brasil.** Universidade de São Paulo, 2010.

PEDROXA, M. M. et al. Aproveitamento Energético De Resíduos Sólidos Urbanos Em Processo De Pirólise. **Revista Brasileira de Energias Renováveis**, v. 6, n. 2, 2017.

PEREIRA, E. R.; DEMARCHI, J. J. A.; BUDIÑO, F. E. L. **Biodigestores- Tecnologia para o manejo de efluentes da pecuária**, 2009.

PINTO, P. H. M. **Tratamento de Manipueira de Fecularia em Biodigestor Anaeróbio para Disposição em Corpo Receptor, Rede Pública ou uso em Fertirrigação.** Universidade Estadual Paulista, 2008.

PISANI JUNIOR, R.; CASTRO, M. C. A. DE; COSTA, A. Á. DA. Development of correlation for estimating the per capita generation rate of urban solid waste in the state of São Paulo: influences of population, per capita income and electric energy consumption. Engenharia **Sanitaria e Ambiental**, v. 23, n. 2, p. 415-424, 29 mar. 2018.

POINT, G. P. DA. **Electricity generation in Isolated Systems: challenges and proposals for increasing the participation of renewable sources based on a multicriteria analysis.** PUC-Rio, 2019.

PRSCS-RMM. **Solid Waste and Selective Collection Plan of the Metropolitan Region of Manaus**, 2017.

RENÓ, F. A. G.; STREB, C. S.; PIUNT, R. C. **Conservation and energy production from solid waste - alternative for two problems: garbage and waste.** (IX Congresso Brasileiro de Energia - CBE, Ed.)Rio de Janeiro: IV Seminário Latino Americano de Energia, 2002

RIBEIRO, L. A. D. S. Making isolated renewable energy systems more reliable. Renewable

Energy. v. 45, p. 221-231, 2012.

ROSA, B. P. et al. Impacts caused in waterways by deactivated controlled landfills in the Municipality of São Paulo, Southeastern Brazil. **Brazilian Journal of Environmental Management and Sustainability**, v. 4, n. 7, p. 63-76, 2017.

SOARES, L. V. et al. **Rate of per capita generation of municipal solid waste for small municipalities of the mesoregion of Sertão Paraiba**, 2015.

SOUZA, J. et al. Urban **waste treatment, energy generation and fertilizer: a perspective for the Vale dos Sinos regionCONGRESSO** INTERNACIONAL DE TECNOLOGIAS PARA O MEIO AMBIENTE, , 2012.

TEIXEIRA, E. N. Resíduos sólidos: minimização e reaproveitamento energético. **Seminário Nacional Sobre Reuso/Reciclagem de Resíduos Sólidos Industriais**, p. 29-31, 2000.

USITRAR. **Carbonization industry , waste processing and power generationS&S** Real Estate Business, , 2017.

VERMA, S. **Anaerobic digestion of biodegradable organics in municipal solid wastes**, 2002.

ZANTA, V. M.; FERREIRA, C. F. A. Gerenciamento integrado de resíduos sólidos. In: **Resíduos Sólidos Urbanos: Aterro Sustentável para Municípios de Pequeno Porte**. 1. ed. Rio de Janeiro: ABES, 2003. v. 1p. 1–18.

Printed by Books on Demand GmbH, Norderstedt / Germany